CREATING YOUR
ECO-FRIENDLY GARDEN

This book is for Bradley and Stacy
in the hope that when they have travelled the wanderlust out of their systems
and decided to put down some roots, they will find inspiration and practical assistance
in these pages. And for everyone with a new garden to create.

CREATING YOUR ECO-FRIENDLY GARDEN

MARY HORSFALL

CSIRO
PUBLISHING

CSIRO PUBLISHING GARDENING GUIDES

National Library of Australia Cataloguing-in-Publication entry

Horsfall, Mary, 1949–
Creating your eco-friendly garden/author, Mary Horsfall.
Collingwood, Vic.: CSIRO Publishing, 2008.

9780643094949 (pbk.)

CSIRO Publishing gardening guides

Includes index.
Bibliography.

Organic gardening – Australia
Gardening – Australia

635.04840994

Published by

CSIRO PUBLISHING
150 Oxford Street (PO Box 1139)
Collingwood VIC 3066
Australia

Telephone: +61 3 9662 7666
Local call: 1300 788 000 (Australia only)
Fax: +61 3 9662 7555
Email: publishing.sales@csiro.au
Web site: www.publish.csiro.au

Front and back cover photos show the development of the author's garden.
Photos by Rodney Horsfall.

Set in 10.5/14 Adobe ITC New Baskerville
Cover and text design by James Kelly
Typeset by Desktop Concepts Pty Ltd, Melbourne
Text printed on paper sourced from well-managed forests
Printed in China by Imago

CONTENTS

ACKNOWLEDGEMENTS

The photos that chart the progress of our garden and bring life to this book are the work of my husband, Rodney Horsfall. His contributions to the garden project included irrigation expertise, organising men with the right equipment for heavy jobs, erecting lattice, creating fence art, bartering a slab of beer for a truckload of mulch and his professional skills for a truckload of rocks, collecting cardboard cartons for sheet mulch, plus unfailing support and encouragement. I might have been the gardener, but it would have been a very different garden without his help. Loving thanks.

INTRODUCTION

The time had come to sell our 40 acre property with its extensive food, native and cottage-type gardens, all developed using organic methods. We had made the decision to move to the country in search of a self-sufficient lifestyle 24 years earlier. We built our own house, kept livestock, grew and preserved much of our food, spun and knitted the fleeces from our flock of coloured sheep, revegetated areas of the property and became adept at numerous other skills of country living. But most of all we gardened. Building on a fascination with native plants and the knowledge gained in a previous garden, we transformed a house paddock of waist-high bracken and rocky infertile soil into a productive, high-biodiversity oasis.

We told ourselves that in our new life in a country town in north-central Victoria we would buy an established house with a garden that we could gradually improve. We didn't want to start another garden from scratch. However, a quick study of the real estate market in the area soon convinced us of the economic sense of buying land, building a new house of our own choosing and starting a garden from scratch.

The land we bought is a double block – 40 × 40 metres – larger than the average suburban block, but smaller than our previous home. The challenges we faced are the same as those facing the tens of thousands of Australians each year who start with a house on a bare or nearly bare block of land.

We decided that environmental values would be built into the garden from the start. We humans too often unthinkingly have a negative environmental impact; we use the earth's resources, leaving nothing but waste in their stead and fail to consider the importance of the natural systems that are essential to our survival. It is within our power to reduce our environmental impact, and creating an eco-friendly garden is a great way to start. Our gardens can continuously improve and replenish the soil, use no damaging chemical fertilisers or pesticides, be water-efficient, incorporate pre-used materials where possible, include native and indigenous plants and be biodiversity-friendly. I am sure our experience of creating an ecogarden from scratch in a country town will help you create an ecogarden, wherever you live – for a more sustainable future.

TAKING STOCK, SETTING GOALS

You have bought a new house, or are in the process of having one built. Maybe you have bought an old house with a minimal garden, one of those all too familiar patch-of-lawn-and-tangled-shrubs-around-the-edges jobs. Having made this major commitment, it is only natural that your thoughts will turn to how you can create a new garden that will enhance your home and complement your lifestyle.

Where to begin can be a puzzle for the inexperienced. Eagerly visiting the nearest nursery and buying any plants that catch your eye is one strategy that might prove successful, if you're lucky. But impulsive purchases are more likely to turn out to be a waste of time and money in the long term. I advise a period of quiet contemplation, gathering information and thinking carefully about what you want before any plants are bought or any earth moved.

Buying a new house is an exciting event in anyone's life, but at the same time the legalities, the myriad decisions to be made, the planning, the packing, leaving friends and family, can also make it a very stressful experience. Take some relaxing time out while the house is being built, or in the period between signing the contract and taking possession of your new home, to think about your new garden.

- Look at gardens around you with new eyes, thinking about what you like and dislike and what plants or features might suit you.
- Wander through public gardens.
- Visit display home villages and notice how the designers have landscaped gardens around the houses.
- Browse through gardening magazines and books.
- Watch television gardening programs.
- Visit open gardens.
- Take photographs of plants or design ideas that appeal to you.

Ideas can come from any number of places. You could start a scrapbook of pictures and plant names and habits. Include in the scrapbook details about garden ornaments, paving types and colours, planting designs and colour schemes that catch your eye. Alternatively, set aside a Manila folder, box, suspension file or drawer for your collected information and inspirations. If you have children, involve them in this process so they will be interested and cooperative when the time comes; they might even want to have

Public gardens like the Canberra Botanic Gardens can provide inspiration for your new garden.

their own patch of ground to develop. If this is not your first garden, you will probably already have lots of ideas. Be aware that what worked in one location might not be so successful elsewhere, so keep a flexible outlook.

Do not buy any plants or other materials at this stage. This is ideas time only.

Take stock of what you have

Will you be starting with bare earth, have lawns already sown, paths and driveway installed, or even a minimal garden planted?

In our case, we had subcontracted the construction of the house, so not only did we start with bare earth around the house-site and a tangle of weeds elsewhere, we also had to tidy and remove the builders' rubble. Some house and land packages come complete with lawn, paths and a few plants. This gives you the opportunity to live in the house while deciding how to proceed without feeling pressured to do something with an unsightly bare block, but don't let what is already there restrict your thinking.

Learn about your soil

The soil is the starting point for your garden and much depends on it, so some basic knowledge about the soil is worth having, even though you might not be planting anything for a while. It is a good thing you can do while your house is being built. Don't be daunted by the scientific look of some of the terms used about soil – acidity, alkalinity and pH – I'll keep it simple.

To help you find out about the soil you will be gardening with, take a trowel and a spade on your next site inspection and dig around in different spots. The most obvious thing you will notice is the soil texture. All soil is made up of particles of sand, silt and clay, and there might be gravel as well. The particles are of varying diameters, with clay particles being the smallest and gravel, if present, the largest. In the majority of cases there will be no gravel, so the sand particles will be the largest. The combination of particles might vary from place to place within every garden. In good garden soil there will also be decomposed organic matter in the topsoil. A desirable garden loam is a friable mixture of different sized particles combined with organic matter, but loams with too much clay or silt can set hard or form a crust.

Texture test

To determine the texture of your soil easily:

- Take a quantity in the palm of your hand – about two tablespoons.
- Gradually moisten it. If it has recently rained, you might not need to add any water.
- Knead it with your hands to thoroughly moisten and roll between palms to make a ball, about the size of a golf ball.
- Now, slowly squash the ball.
- If you couldn't even roll a ball in the first place, the soil is coarse sand. You will feel, and maybe hear, the gritty texture on your hands.
- The more easily the ball crumbles under pressure, the higher the sand content.
- If the ball splits but does not fall apart and has a silky feel, it is silty and/or loamy.
- If the ball flattens, but stays more or less in one piece and feels sticky or elastic (like plasticine), it is high in clay.
- Pull and stretch a ball of clayey soil between your fingers to form a ribbon. The longer the ribbon, the more clay the soil contains.

Of course, there are proportions of sand, silt, clay and organic matter, but a general idea of the kind of soil you will be working with is all you need. To see the proportions of sand, silt and clay in your soil take the same soil sample you rolled into a ball and place it in a screw-top jar half-full of water. Put the lid on and shake the jar vigorously. Let it settle. The soil will have separated into layers with coarse sand or gravel (if any) on the bottom; on top of this will be any finer sand, then silt, then clay on the top.

Soil structure

If texture refers to the size of the soil mineral particles, structure refers to the way the mineral particles and any organic matter are bonded together into crumbs or aggregates. The aggregates cluster together in distinctive ways for different soils, leaving spaces between for air and water. We have all seen a clod of clay. This is made up of clay particles bonded together so firmly that it holds its shape even when rained on or dug. This is poor soil structure that does not allow water, air or plant roots to penetrate it. Another type of poor soil structure is very sandy soil with little clay or organic matter. This has a different, more open aggregate pattern that cannot retain water and nutrients.

Soil can be improved

The best type of soil for gardening is friable loam (a balanced mixture of soil particles) with plenty of organic matter in it. Many plants will not thrive in extremes of either sandy or clay soils, so the soil type will influence what plants you choose. Do not be unduly concerned, though. The good news is that most soils can be improved quite easily.

Remember:

- Sandy soil is very free-draining, but low in nutrients and organic matter. It will consume vast amounts of water, yet not hold it to support plant growth. Added nutrients are likely to be quickly leached away.
- Clay soil holds nutrients well and can hold water too well, resulting in drainage problems. Clod structure will prevent air, water and plant roots from penetrating.
- Clay soil that is slippery or sticky when wet, that dries to a hard crust, becomes very muddy when it rains and then drains slowly will often benefit from the addition of gypsum. Use about a half to one kilogram of gypsum per square metre. There are liquid preparations to add to clay soil to improve its texture and drainage ability, either a simple liquid gypsum or a mixture with fertiliser and other soil conditioners. These same preparations are also said to improve the water-holding capacity of sandy soils.
- Sandy soil can have clay worked into it. You will need about a quarter the volume of clay to sand.
- The addition of generous amounts of organic matter in the form of decomposed manures and mulches will, in time, improve the texture, structure and nutrient availability of any soil.

Soil profile

To find out more about your soil, use a spade to dig up a slice of soil about the width and depth of the spade. You should be able to see a change in colour and texture from top to bottom of the profile.

- A layer of decomposing mulch, dark, sweet-smelling topsoil and roots visible all the way through the profile are all indicators of fertile, well-drained soil.
- If the block of soil is hard to break up when dry, has a sour smell when wet and looks blue or grey and yellowish grey, this probably indicates poor drainage and heavy clay. Such soil could be subject to periodic waterlogging.
- If the soil falls apart as soon as the spade is removed, it is most likely to have a high sand content.
- The presence of earthworms or their tunnels is an indicator of good aeration, drainage and fertility.
- Beetles, grubs, spiders, ants and other soil biota that you might not be able to identify indicate a healthy level of biodiversity in the soil.
- This digging around will also show you any places builders' rubble has been buried, perhaps under a load of innocuous-looking topsoil. Depending on the amount and type, buried building materials can prevent proper development of plant roots and can even release substances that will inhibit plant growth. The simplest solution is to plan on building up raised no-dig beds over any such areas, or you might prefer to hire a contractor to remove the rubble and begin again with new topsoil.

Your situation could allow you access to the building site at the foundations stage, so you might not need to do any digging around. Examine the holes dug for the footings to see the soil profile, discussed above. A convenient fall of rain will allow you to check the soil's texture and drainage. If the footings fill with water that seeps straight away, the soil is sandy; the sides of the footings will crumble easily in this case. If the water stays in the holes for days, there is a lot of clay in the soil.

Our soil is mostly heavy clay, with very poor drainage in some spots, so this had a big influence on our choice of plant species and our method of gardening. I decided very early on that most garden beds would be raised and built on top of the existing soil and have lots of organic matter added.

Soil pH

There are people who garden all their lives without ever testing their soil's pH. I have met some of these people and their gardens are usually thriving, but then they might say that their azaleas are struggling, or just won't grow, or that their citrus trees have a mysterious complaint. A simple pH test might have shown them that the soil was neutral or slightly alkaline and needed some adjustment to bring it into the slightly acid range preferred by these plants. Testing for pH is a doddle to do and can give you valuable information about your soil and the mulches you will be using later on.

The soil pH refers to how acid or alkaline it is. The pH scale ranges from 1 to 14. At one end of the scale pH 1 is the extreme of acidity,

while at the other end pH 14 is the extreme of alkalinity, and in the middle, pH 7 is neutral. Many plants prefer a slightly acid soil in the range of pH 6.2 to 6.5, though most will grow perfectly well from pH 5.5 to 7.5. Some plants have a definite preference for either an acid or an alkaline environment and will not thrive if the soil pH is too far from their preferred range. So, soil pH has a big influence on plant choice.

Another important reason for finding out the pH of your soil is that at different pH levels some nutrients can be unavailable to plants and some can be taken up in toxic levels. If the pH is too far into either extreme, most plants will not thrive.

You don't need to be concerned about the science behind soil pH, just buy a cheap pH test kit, one of the colour-coded ones, from a nursery and test some samples from around your block. These kits might not be 100% accurate, but are very easy to use and give you a good enough idea of what you have.

Adjusting the pH

Do this once you are actually ready to start gardening. To raise the pH of acid soil, apply dolomite, lime, wood ash and alkaline mulches. To lower pH of alkaline soil, apply sulphur or iron sulphate and use acidic mulches such as pine needles, sawdust and tanbark.

Mulch and pH

When it comes time to buy mulching materials, it is sensible to purchase a small amount, or obtain a sample, and test the pH before using it. You will want to avoid using mulch that will exacerbate an existing pH imbalance. I have been quite surprised at the pH of some mulches I tested:

- Sugar cane – 5.5 – acid.
- Mixed packaged manure – 9 – alkaline.
- Stable sawdust from a local industry – 8.5 – alkaline.
- Packaged soil improver and mulch – 5.5 – acid.

The texture of many mulches makes them difficult to test with the colour-coded kit, which is designed to use with soil mixed with the liquid solution provided. Some can be softened with a little water, allowed to dry and then crushed to a powder. Some can be pulverised to a powdery consistency in a vitamiser.

Our soil is quite acidic with a pH of 5.5. I wasn't overly concerned about this because most of the plants I wanted to grow prefer an acid environment. I was intrigued to find, though, that when I mixed some soil with a variety of mulches I was using and again tested the pH, the mixture had a pH of 7, perfectly neutral. Though the ratio of mulch to soil will be quite different in the real garden, this little experiment illustrates the role mulch can play in stabilising soil pH.

Study your surroundings

By the time your house is at lock-up stage you need to become aware of the surroundings that will provide the backdrop for your new garden.

- Get your compass bearings and notice where the sun rises and sets, where the

A flowering pear in a neighbour's garden lends its charm to my garden as well.

shade of buildings, fences and neighbouring trees falls, and at what times. Also take note of any areas that are exposed to the sun for much of the day; these will be ideal for hardy native plants or xeroscape gardens. (A xeroscape garden is a botanical term for a garden of plants that will live through drought conditions. They often consist of hardy native plants mulched with wood chips or succulents and cacti surrounded by pebbles.)

- Neighbouring buildings, especially multi-storied ones, can have a big impact on your garden. They might shade large areas at some times of day, give protection from strong winds, be a source of radiant heat or create a rain shadow that prevents rainfall from reaching some areas.
- Trees on neighbouring blocks might overhang your fences and their roots intrude into your gardening area, or they might be an attractive feature that can complement your own garden design.

- Roads, driveways and paths can provide radiant heat and rainwater runoff to adjacent garden beds.
- Notice the direction strong winds come from and plan how to use buildings, fences or windbreak plants to modify them if necessary.
- Find the areas that are well protected from wind and plan to use these for vegetable gardens or more delicate plants, or as tranquil retreat zones.
- You might be fortunate enough to have a view or pleasant outlook of something other than rooftops, even in an urban setting. Consider how to design your garden to frame it so it remains a focal point.
- Where neighbouring sheds, roofs or other unsightly objects will be in your view, plan how to hide them. This could be with quick-growing shrubs, lattice that will support a quick-growing vine, a pergola or gazebo.
- Talk to your new neighbours, as they can be invaluable sources of information, especially if they have been in the area for

Surrounding fences provide a protected microclimate for the vegetable garden.

a while. Several people mentioned to us that termites were a problem. One neighbour even said that she had seen garden stakes eaten out by termites within two years. Forewarned, we decided not to have garden beds, which could give ready access to the pests, abutting the house.

Climate

If your new home is in an area you have lived in for a while, you will already be familiar with the climate and have a fair idea of what plants will thrive there. If it is in a region you are unfamiliar with, it is wise to take some time to look around other gardens in the area, talk to neighbours and local nursery proprietors.

Within your garden there will be a variety of microclimates resulting from the placement of structures (your own and those of your neighbours) and surrounding plants. Use these to your advantage when trying to grow plants less suited to the general climate of the region. Some years ago, for example, the flamboyant beauty of tropical hibiscus plants captivated me. Knowing there was no possibility they would thrive out in my windswept, exposed garden in inland Victoria, I planted them in pots and placed them against a sheltered north-facing stone wall. Every two or three years I divide and repot them and give many root divisions away. A city gardener I know in Melbourne was able to grow bananas in her garden by taking advantage of the radiant heat from the brick wall of a neighbouring block of flats.

If there is something special to you that would not normally grow in the climate of the region, now is the time to consider how to take advantage of an existing microclimate, or how to create one to suit your favoured plants.

Existing plants or features

You might already have some plants on your block. In our case we had two mature red gums on the east side and several trees comprising a couple of different species of eucalypts, not terribly old but still a good size, on the nature strip. As well as being features in their own right, the gum trees provide dappled shade for the native shrubs we planted beneath and nearby. There were also a kurrajong tree, two unidentified small trees or shrubs, and several self-sown thorny plum trees. We knew we could find a creative use for some slabs of red gum trunk that we inherited.

The plants you have inherited might not be to your liking. Do not feel inhibited about getting rid of them. You might know someone who likes them, or maybe they would just look

A mature red gum that was an existing feature adds height to a new area of native garden.

better in a different spot. Be ruthless also with plants that will grow too large for your garden. Most new gardens are too small to waste space on plants you dislike or that are not suitable for the location. We removed the thorny plum trees, and we transplanted the unidentified young trees from the nature strip to a spot at the back where they would screen a neighbour's shed until we could identify them and decide whether or not they would remain (they were privets, and they didn't).

Your block might be flat. Consider landscaping to introduce a slope, terraces or mounds. A steeply sloping block presents a different challenge and might suit you better terraced. A poorly drained spot could become a pond or bog garden. A stony area or rocky outcrop can be part of a native or succulent garden feature.

We have a low area along the eastern side, where the red gums are, that is a natural drainage channel during heavy rainfall. Instead of regarding it as a problem, we decided to turn it into a feature. It is now a seasonal pond and stony creek-bed admired by all, including many locals who warned us of the 'flooding problem' we would have.

You might already have a shed, fences or other structures in the garden. Unless they are very unsightly or you have an unlimited budget, these should be regarded as permanent fixtures. You need to decide how to incorporate them in your garden design, how to hide, disguise or feature them, depending on the situation.

Existing paths will also usually stay, unless they are of pine bark or similar soft material.

Find suppliers

You are going to be doing a lot of business with local nurseries, hardware stores, landscape suppliers, contractors and tradesmen. Find out where they are, the range of goods they sell and how prices compare so when the time comes to buy plants and other garden supplies you know where to go. Ads in gardening and environment magazines and local newspapers are very useful. Cut them out and add them to your scrapbook.

As soon as it rained we had native ducks visiting the new pond.

Freebies and bargains

As you explore the area keep a lookout for possible sources of free or cheap materials. Thick cardboard is a good sheet mulch to control weeds. Local businesses will probably be happy for you to help yourself. Are there riding stables or horse studs that get rid of their animal bedding? Farms nearby could supply straw, spoiled hay or animal manures. Tree lopping contractors might be willing to dump a load of mulch at your place. Check your council's green waste policy. In our Shire, shredded, partially decomposed mulch is available free if you collect it yourself from the tip.

Second-hand timber yards, garage sales, markets and clearing sales are possible sources of materials for paving, edging, garden ornaments, or timber for building pergolas and other garden structures.

Setting goals

By now your mind is probably buzzing with ideas. The challenge is to put all the information and possibilities into some sort of order so you can make a workable plan. It is time to decide what you are aiming for, what you want from your garden. I decided that I was aiming for:

- Beauty – an aesthetically pleasing environment to live in.
- Biodiversity – a variety of indigenous, native and exotic plants to attract native birds and other fauna.
- Bounty – a productive food garden within the space constraints of the block.

All of this had to be achieved on a tight budget, without using garden chemicals. It must require minimal digging and maintenance once established, and be water-efficient. You might have different priorities. Perhaps you would like to incorporate:

- an entertainment area
- a water feature
- a children's playground
- a pet enclosure
- a perfumed garden

- a swimming pool
- a theme garden
- a herb garden
- an existing collection of plants, statuary or other garden ornamentation.

Maybe you want to:

- have a no-mow garden
- specialise in a particular species or type of plant

- blend in with the neighbours
- be different from the neighbours.

To help with setting your goals, make a list of must-haves and would-like-to haves. How does your list measure up against the realities of your situation, those points you noted earlier? Is there space for that swimming pool, for example, and does the climate of the region allow frequent enough use to justify the expense, maintenance and water consumption?

One of our major goals in creating our new garden, as it was with our previous farm garden, was to encourage biodiversity. The next chapter outlines our reasons for making this a focus right from the planning stage, explains why it is relevant to everyone developing a garden, and gives you some ideas about how to incorporate biodiversity values in your new garden.

THE BIODIVERSITY CONNECTION

Biodiversity loss is one of the most crucial environmental issues facing the world. All other environmental problems – climate change, salinity, acidification, soil degradation, pollution, water shortages and deforestation – contribute to biodiversity loss. Some environmentalists predict the loss of 50% of the world's species within the next few decades. Here in Australia a 2002 report identified 2891 threatened ecosystems and a total of 1595 threatened species. Vegetation-clearing and increasing fragmentation of remnant vegetation were cited as the most significant threats to species and ecosystems.

You might be wondering what this has to do with your new garden.

To be brutally frank, our new homes are part of the problem. My new home and yours are built on areas of land that were once ecosystems of native flora that provided habitat for a range of fauna. Inevitably, the homes built to house a growing human population cause habitat destruction and fragmentation and displace the indigenous flora and fauna. It seems only fair then, and could in fact be crucial, that in planning our gardens we keep this in mind and try to incorporate plants, features and gardening methods that are biodiversity-friendly.

In most cases it will not be possible to re-create the biodiversity of the original ecosystem, or even anything very like it. What you will be creating is an area of artificial biodiversity. However, given the enormity of the problems associated with biodiversity loss, I believe that artificial biodiversity is better than severely reduced biodiversity. I think we need to be realistic enough to accept that artificial biodiversity, at least in urban areas, is the best we can do.

A variety of grasses, shrubs and rocks chosen to encourage biodiversity by providing food and habitat for native fauna.

In developing your new garden constantly ask yourself, '*Will what I'm doing now enhance biodiversity or reduce it?*'

More than koalas and bilbies

It is important to recognise that biodiversity encompasses much more than bilbies and koalas. It includes every living thing we see around us, including ourselves, as well as myriad micro-organisms we can't see but which are crucial to our very existence. The life forms of an area together with its non-living features – hills, mountains, valleys, hollows, lakes, coastlines, dunes, rocks –

A pleasant surprise early in our garden development.

create an ecosystem. All our gardens are mini-ecosystems within the greater ecosystem of our region.

No matter how small the garden, there are many ways to enhance biodiversity and reduce our own damaging environmental impact. Our gardens can become part of a gigantic green tapestry threading through the towns and suburbs, and linking up with parks, reserves, creek and river banks, and roadside plantings. In this way we create a biolink giving protection, habitat and food to a range of wildlife either passing through or becoming permanent residents. Birds, insects, frogs and lizards are the most likely visitors to biodiversity-friendly gardens, but you might be surprised at what turns up on your doorstep.

Narrow areas along a footpath in Canberra are planted with native shrubs and link with a nearby park.

In our case, the existing trees on our block and nature strip already link up with others scattered all along the roadside, right down to a creek reserve about a kilometre away. The garden was barely started when we were thrilled to be visited by a koala in the gum trees and several species of ducks on the pond, as well as a variety of other bird life. The understorey of shrubs we intended to plant would attract an even greater range of birds and the insects they need to feed on.

Planting for biodiversity

We choose most of our garden plants for the beauty of their flowers. Giving pleasure to humans, though, is not the reason plants put so much of their energy into flower production. Plants ensure their successful reproduction by producing flowers to attract pollinators. The pollinators might be insects, birds or small mammals which, in their efforts to obtain the energy-rich food of nectar or pollen produced by the flowers, transfer pollen from the anthers of one flower to the stigma of another, ensuring fertilisation and seed set.

Over aeons of shared evolution, flowers have developed a variety of ways to attract the pollinators on which they depend and pollinators have often developed preferences for particular food sources. Some pollinators are attracted by specific flower colours, others by perfume, the shape of the flower or its position on the plant.

To attract pollinators to our gardens it is necessary to plant a variety of flowering species of different colours, shapes and perfumes. Gardens based on a limited colour scheme or type of plant, while they might conform to a short-term human ideal of desirability, will not appeal to a diversity of pollinators and will be low in biodiversity value. There is a place for the architectural elements of mondo grass, plaited ficus trees

It wasn't long before many of our new plants were flowering, attracting numerous pollinators.

and box hedges, but for anyone who values biodiversity it is a very limited place.

The importance of pollinators

Why do we want to attract pollinators in the first place? The broadest answer is that many of our agricultural crops are dependent on them. The creatures that provide pollination services of immense value to mankind are increasingly threatened by habitat destruction and fragmentation, and by agricultural chemicals. Our gardens can act as reservoirs for pollinators, thus ensuring their survival and the survival of our own food sources. Similarly, anyone who grows their own vegetables and fruit, or who wants their flowers to set seed for future propagation, needs pollinators in their garden.

A bird-eat-bug world

There is another important reason to attract pollinators that has little to do with their pollination services and a lot to do with the fact that they are part of a food chain. In the natural world everything eats something else in order to survive. Those same insects that pollinate our flowers can also prey on or parasitise a range of pest insects and help keep them under control. They are in turn prey for birds, frogs or lizards. The honeyeaters, for example, that pollinate our flowers while feeding on the nectar within them, also eat insects from under the bark of trees and might themselves become prey for larger birds. Lizards scuttling in the mulch are devouring a range of garden pests and might end up being eaten by magpies.

The pretty gold dust wattle (*Acacia acinacea*) – a compact shrub indigenous to our area.

Variety is crucial

The key element to attracting a diverse range of insects, birds and other fauna to our gardens is variety of flowering plants. It is thus important too that there are some plants in flower at all times of the year. A layered

Layered planting and different flowers throughout the year offer maximum food and habitat for resident and visiting fauna.

planting design that incorporates trees and shrubs of different mature heights as well as low-growing shrubs and ground covers will give the maximum choice of habitat for birds and other fauna.

Though some native fauna species have very specific food and habitat requirements, others are generalists and will adapt readily to a range of garden situations. Native plants, especially those indigenous to the area, are likely to offer food and habitat to attract local fauna, but many exotic plants can fill this role as well, and in some cases might be preferred.

Seeking indigenous species

In most areas now the indigenous plant species are unlikely to be well represented or readily available, so anyone wanting to plant them will need to do some searching. Approach local nurseries, councils, water authorities and environment groups for any information or publications they might have.

Organisations such as Greening Australia or the Societies for Growing Australian Plants in each state could be helpful. The Flora for Fauna program has a useful website: www.floraforfauna.com.au.

In some locations there are organised programs to plant indigenous species that might be locally threatened, or that are vital food or habitat for locally threatened birds or butterflies. Why not participate and make your garden part of the solution?

Plants that attract

Plant any indigenous species you can track down in order to attract indigenous fauna. Many other native species and exotics are suitable as well. The size of your garden will limit plant choice to some extent, but many of the plants recommended for attracting birds and insects have dwarf and miniature forms that can be included in most small gardens. Plants marked * are suitable for small spaces or have small forms readily available.

Banksias are a favourite source of food for many birds and insects.

Honeyeaters and other nectar-feeding birds can be attracted by:

- grevilleas*
- banksias*
- kangaroo paws*
- callistemons*
- salvias*
- red-hot pokers*
- penstemons*
- yuccas
- buddleias[#]
- strelizias
- proteas
- camellias*
- wattles*
- correas*
- epacris*
- paperbarks
- ericas*
- flowering gums (many varieties are suitable for medium sized gardens).

[#]Note that buddleias are classified as environmental weeds in some areas (see 'Plants to avoid', page 17).

Seed-eating birds can be attracted by:

- banksias*
- conifers
- tea trees*
- wattles*
- casuarinas
- eucalypts
- crepe myrtles
- silver birches
- grasses (native types, not lawn grasses).*

I have been delighted to see both crimson and eastern rosellas feasting on crepe myrtle and silver birch seeds during winter when there is not much else around for them.

Casuarinas come in many forms, all loved by seed-eating birds.

Lilly pilly berries.

Fruit-eating birds can be attracted by:

- lilly pilly*
- bangalow palm
- wild cherry
- blueberry ash
- flax lilies.*

Nectar-feeding and pollen-collecting insects can be attracted by most of the same flowers as nectar-feeding birds, as well as:

- erigeron*#
- alyssum*#
- borage*

- bergamot*
- tea tree*
- parsley and carrot plants in flower*
- any open daisy-like flowers.*

#Both erigeron and alyssum have weed potential in some areas.

Attracting butterflies

Many of the above-named plants will attract butterflies to your garden, but the following small native species will be especially

Don't forget, grasses are good for seed-eating birds.

Carrot in flower is adored by many pollen-collecting insects.

Native daisy flowers are great for attracting butterflies.

beneficial and are colourful understorey plants as well as being ideal for small areas:

- *Brachyscome diversifolia* (tall daisy) and *B. multifida* (cut-leaf daisy)
- *Bracteantha viscosa* (shiny everlasting)
- *Brunonia australis* (blue pincushion)
- *Chrysocephalum apiculatum* (common everlasting) and *C. semipapposum* (clustered everlasting)
- *Craspedia glauca* (common Billy-buttons)
- *Helichrysum scorpioides* (button everlasting)
- *Leucochrysum albicans* (hoary sunray)
- *Indigofera australis* (austral indigo)

Plants to avoid

One of the threats to biodiversity throughout Australia comes from environmental weeds – garden or pasture escapees that have proliferated to such an extent that they have changed ecosystems to the detriment of the indigenous flora and fauna. Some of the worst culprits are well known and easily

recognised and you wouldn't want to plant them in your garden anyway. These include Paterson's curse and blackberries. Did you realise, though, that popular plants such as agapanthus, Cootamundra wattle, lantana and privet are classified as environmental weeds in some areas? Find out the environmental weeds of your area from your council or environment group and avoid planting them. The Weeds Australia website at www.weeds.org.au lists noxious weeds Australia-wide. A list of invasive garden plants can be found on the WWF Australia website, www.wwf.org.au, and the Victorian Government's Department of Primary Industries website: www.dpi.vic.gov.au.

Environmental weeds include some native plants that have become pests through having been planted outside their normal range, and having propagated themselves so successfully in the new habitat that they have displaced indigenous species.

To lawn or not to lawn

Traditional lawns are monocultures. As such they are anathema to biodiversity. They require huge amounts of water and lots of maintenance, which uses fossil fuel and causes noise and air pollution. Too often, the fertiliser needed to keep them green and lush runs off or leaches through the soil and ends up in waterways, causing toxic blue-green algal blooms and benefiting aquatic environmental weeds. That said, lawns do have their uses. They are great for young children to play on, frame garden beds in an attractive way, and many find a smooth green sward extremely refreshing to the eyes, especially in summer.

For an in-depth discussion on how to grow a sustainable lawn anywhere in Australia, even when water restrictions are severe, see Chapter 12 in Kevin Handreck's *Good Gardens with Less Water*.

An intriguing theory

Perhaps our illogical devotion to the-greenest-lawn-in-the-street mentality can be explained by a theory I read recently. Apparently, when given a choice of landscapes, most people choose one with a savannah (grassy) foreground blending back to an open treed background. The theory suggests that it is because of a tribal memory of the African landscape where humankind originated. If this is the case, our desire to include large grassed areas in our gardens is not a free choice at all, but driven by deeper needs.

A multitude of choices

Whatever the reason for our choice, it is time to think it through more carefully. On any logical basis the green lawn is supremely unsuited to most Australian conditions. When planning your new garden, consider the following suggestions.

- Reduce or eliminate traditional lawns.
- Choose grass varieties that need little water and fertiliser once established. These include some named varieties of soft-leaf buffalo and Arid Smartgrass. Warm-season grasses such as couch will often grow well with little summer watering; however, they tend to go yellow over winter and can be invasive.
- Select native grasses such as Hume wallaby grass, windmill grass, Queensland blue

Our herb lawn is spreading nicely and the flowers are swarming with bees.

grass, red-leg grass or meadow rice grass. These are drought-tolerant once established and require little or no fertiliser.

- Just mow the grasses and weeds that grow naturally on the block and do not water them in summer. You will probably find, as we have, that there are enough summer-growing species present to give a greenish tinge, though it is certainly not lush, throughout most of the year without extra water. Because of their different growth habits, frequent mowing encourages grass growth and discourages broadleaf weeds.
- Choose lawn alternatives such as chamomile, pennyroyal, native violet, kidney weed, thyme and lippia. These still need water over the hottest months, but possibly less than most grasses. They can be mown occasionally and need little if any fertiliser. A mixture of alternatives is more in keeping with biodiversity values and the herbs, when in flower, will be teeming with pollinating insects.

- There are good synthetics available which are worth investigating, especially for anyone with a small lawn area.

Designing for biodiversity

To survive and thrive in our gardens the insects, birds and other fauna we wish to attract need more than plants. Fortunately, most of their other needs can be met by the inclusion of features that are attractive to the garden's human inhabitants as well.

Water

Nothing adds to a garden's ambience as much as a water feature or two, and every garden, no matter how small, can include water in some form. A very small area, even a balcony, can include a birdbath or some glazed bowls of water. There are numerous sizes and styles of pre-formed ponds available from nurseries, or enterprising and energetic gardeners can construct their own.

A good pond ecosystem, with rocks for birds to land on and vegetation in the water to shelter and feed aquatic life.

A SIMPLY PERFECT POND

The simplest option for including a pond in your garden is to buy a pre-formed fibreglass or plastic shape and have a hole dug to suit it. Allow sufficient depth to spread a layer of sand for the pond to nestle into. These pre-formed ponds are available in a wide range of sizes and shapes. Remember that if you want to stock your pond with fish it must have a minimum depth of 80 cm. Ideally, you will have deep water (over 2040 cm), shallow water (over 60 cm) and a bog zone (about 10 cm).

Water in the garden is essential to birds and to many insects as well. Lizards need a drink from time to time, and frogs will take up residence if a suitable pond is available. Make sure that any bowls of water are shallow so birds and other creatures cannot fall in and drown. Try to find a suitable rock, one that projects slightly above the water level, to place in the bowl so birds or lizards that do fall in have a safe exit.

A pond suitable for frogs needs to have both shallow and deep areas. There should be water plants and surrounding foliage to protect frogs when they are out of the water. Overhanging foliage will both shade the water and drop leaves into it; the algae from decomposing leaves are essential food for tadpoles. A floating piece of wood on top of the pond will provide a safe drinking platform for birds and could save any that fall in from drowning. Frogs and fish in the one pond are often incompatible, so if you want to attract frogs to breed in your pond you will not be able to keep fish that would eat the spawn and tadpoles. Some native fish to try in a larger

Rocks have a multitude of uses in the garden and are attractive features in themselves.

pond include native pygmy perch, Murray River rainbow fish and common jollytail.

Rocks

Rocks, whether fake or real, have many uses in a biodiversity garden. Used as garden edging they provide nooks and crannies to harbour a variety of insects, beetles and lizards. Earthworms and other soil biota like to burrow under them where it is often damp and shaded. Featured in a clump or two around the garden, rocks moderate the climate, radiating stored daytime heat during the night and deflecting breezes. Clumps can be arranged so as to leave small secret spaces between them to shelter lizards. Lizards also enjoy basking on rocks to soak up the warmth, and frogs shelter under them on the shady damp side. Rocks placed beside a pond will shelter frogs and give birds a landing place for bathing and drinking.

Be aware that the nooks and crannies that provide habitat for desirable lizards and frogs are also attractive to undesirable garden inhabitants such as slugs and snails. Use this knowledge to your advantage and place snail traps or baits in these spots.

Nesting sites

Birds need somewhere to nest, and for many Australian bird species, and numerous other native fauna as well, this means tree hollows. A hollow large enough for even the smallest bird to nest in can take over 100 years to form. Larger hollows take from 150 to 200 years. The increasing shortage of large trees of this age, along with other habitat pressures, is causing many bird species to decline dramatically in number. If your garden has a suitable predator-safe spot, you can erect a

Gaps around rocks are habitat for many small creatures.

bird nesting box or two. Advice about sizes and shapes for different species and where to locate boxes is available from the Birds Australia website: www.birdsaustralia.com.au. Try also the Bird Observers Club of Australia: www.birdobservers.org.au.

If you do erect nesting boxes, monitor them closely. They could attract unwanted aggressive species such as noisy miners or whatever other bird is a pest in your area. If this happens, remove the box.

Many branch-nesting birds will appreciate a quiet, scrubby overgrown corner to nest in. If the foliage is on the prickly side, like some grevilleas, banksias and callistemons, the birds and their young will be safer from predators. This area will also be great insect habitat. There are gardeners who would consider a scrubby area like this unattractive and inappropriate for their style of garden, but a little imagination and creativity should make it possible for most gardens to contain

BULLY BIRDS

Just as some plants, even native species, become environmental weeds, some animals thrive in our human-created ecosystems to the detriment of others. Aggressive bird species such as wattle birds, currawongs, noisy miners and crows are likely to take over our gardens, driving out and preying upon smaller, less vigorous species. Does this mean we shouldn't create bird-friendly gardens at all? I think the best we can do is provide a range of habitat niches so the smaller birds that might be attracted to visit have refuge areas where they can feed and drink in relative safety.

such a wildlife haven. Screen it from the rest of the garden by an attractive fence or hedge; hide it behind a shed or garage; or put up a sign reading: '*Quiet, Birds Sleeping Here*', or '*Beware, Wild Insects*'.

It is up to us to make sure birds and other fauna attracted to our gardens are safe from domestic pets.

Organics for biodiversity

The driving force of biodiversity consists of bugs and beetles and creepy-crawlies and unimaginable millions of micro-organisms, not the megafauna and the cute and cuddly creatures that usually grab media attention. These diligent but largely unseen workers perform many ecosystem services that include: maintaining the health of our air and water, breaking down pollutants and decomposing wastes into nutrients that improve the soil and feed all our crop and garden plants.

Soil biota

Healthy soil is teeming with life, most of it beneficial to our plants. Within the soil and the mulch live such visible life forms as springtails, beetles, earthworms, spiders, ants, mites and termites. Microscopic organisms such as bacteria, protozoa, fungi, viruses and algae number in the hundreds of millions for every single gram of soil.

All these soil and surface biota are responsible for generating and maintaining soil fertility. Just one of the significant benefits of their complex interactions is the process of decay – they act together to break down the mulch into new soil and nutrients to fuel new growth. In addition, they aerate the soil, create tunnels that allow water and nutrients to seep below the surface to the plant root zone, neutralise soil pollutants, grow in association with plant roots to enable them to reach further into the soil for nutrients, and increase disease resistance in plants. The addition of beetles, springtails or termites to soil has been shown to speed decomposition and enhance plant growth; their removal has the opposite effect.

Much of the beneficial soil biota is killed by garden chemicals and artificial fertilisers, so avoid them and use organic gardening practices to enhance soil health. Make your own compost from kitchen and garden waste, and use animal manures, vermicast, seaweed products and rock dust as fertilisers.

Dig as little as possible. Digging disrupts the balance of micro-organisms, can damage soil structure and cause nitrogen loss from the soil. Use mulch and no-dig beds to provide

the organic matter that encourages beneficial micro-organisms.

Pest control

Organic gardening practices also ensure that your garden has populations of birds, frogs, lizards and beneficial insects to keep pest insects under control without the need for poisonous sprays. It might take two or three years before your garden has developed a healthy flux of biodiversity so there are nearly always enough predators to

- controlling pests by maintaining soil health, encouraging natural pest–predator relation-ships and strategically utilising low-impact sprays. See Chapter 8 for more details.

Many products at the nursery, especially fertilisers, will be called 'organic' somewhere on the label. This does not necessarily mean they contain no synthetic chemicals. Usually all it means is that the contents consist of once-living material. If you want to be sure products contain no chemical residues and were produced using the standards mentioned above, look for an accredited certification logo such as Biological Farmers Australia (BFA) or National Association for Sustainable Agriculture Australia (NASAA).

I do not worry too much about obtaining organically certified manures, for example, because the decomposition process manure goes through before I buy it should already have neutralised any undesirable agricultural residues.

keep pest numbers in check. Any temporary pest outbreaks can be controlled

ACTION PLAN
- Find out about indigenous plants.
- Choose plant species to encourage birds and insects.
- Avoid planting environmental weeds.
- Think about lawns and alternatives to them.
- Consider design features to encourage biodiversity.
- Commit to organic gardening methods.
- Encourage soil biota.

by benign methods such as squashing, hosing, or such relatively safe sprays as pyrethrum. There are many good books about safe pest control, but the most important starting point is creating a biodiversity-friendly ecogarden. See Chapter 8 for more details.

By now you will realise the importance of biodiversity conservation and be aware of measures you can take to make your garden biodiversity-friendly. It is time to incorporate this knowledge into your plan.

PERFECTING A PLAN

The style of garden you would like to have will help determine much of the planning described in this chapter. What has caught your eye when looking at books and magazines or visiting gardens? You might already have inspirational pictures in a scrapbook.

Your garden style will also need to match your family's lifestyle. A choice of formal sculptured beds and smooth lawns is asking for conflict if you have an active family of young cricket and football stars and their bouncy, bone-burying pets.

Do you like a formal style or a more casual style? Do native plants appeal to you? I hope by now you are convinced of the beauty and biodiversity value of natives and full of ideas about how to incorporate them. What about cottage, Mediterranean, Mexican or Balinese styles? Of course, there is no reason why you can't have different styles in different areas or a joyful eclectic mixture if that is what suits you. You will most likely be living in this garden for a considerable time, so the worst choice you can make is to design a garden that suits somebody else's idea of good taste, fashion or desirability. This is *your* garden. Adapt ideas from anywhere you find them, but ultimately *you* have to be happy with your garden. With our emphasis on native and indigenous plants and choice of terracotta red and Mediterranean blue colours for paving and accessories, I think the term that best describes our garden style is Aussie meets Mediterranean.

The angles of house walls, fences and other structures will naturally define a number of different areas or potential plots within the garden, so once the house, shed, fences and any other major structures are in place, your planning can become more concrete. If there are no sheds, fences or other structures, and you want them incorporated, you must decide where they will go so they are both practical and aesthetically pleasing.

You might have moved in by this stage. The conventional approach is to draw up a plan.

My term for our garden style: Aussie meets Mediterranean.

By the time our house was built to lock-up stage I found myself dividing the block into patches or plots, defined by the shape of the house, the position of the fences and the block's topography, and forming mental pictures of what I wanted to do in each plot.

Plot planning

Here is how my initial thoughts began to take shape.

Succulent plot

I thought the long narrow area between house and fence on the western side would be a very hot area where a pebbled succulent garden would do well and require little care long term. The necessary, but decidedly unattractive utilities – hot water service, heating unit and wall-mounted clothesline – were located here, so access to them would need to be maintained. I visualised a vine-covered lattice attached to the fence to make it higher and form a screen from the close neighbour on that side. The concept for this plot was adapted from similar effects used in many display homes we had visited.

Most houses have a similar long, narrow area, not always on the western side. If your narrow plot is on the shady, south side, it could be a perfect place for a fernery. Play with some ideas that suit your own situation and gardening goals.

South nature strip plot

The front of the house, facing the road, looks south. During building we noticed how it was buffeted by strong southerly winds. The house design meant that this area had to be given over mainly to driveway access, but we did have a wide nature strip with some large gum trees already on it. I could see a cluster of native shrubs under the gum trees to both filter the wind and give us some privacy from passing traffic. Plants in this plot would need to be quick growing, hardy and bushy. Once established, I hoped they would survive without extra watering. Some councils have definite requirements about what can and cannot be planted on nature strips. Check before making any plans for this area.

Pond and creek-bed plot

Everyone we spoke to locally was concerned that this low-lying floodway area would be a big problem for us. We regarded it as an attraction of the block and were excited by the prospect of turning it into a feature. I don't know where this concept came from, but the first time it rained after we bought the block and we could see where the water flowed and how long it stayed, the picture just appeared in my head fully formed.

The lowest area would be deepened to form a natural seasonal pond between the large red gums. The overflow of the pond would follow the existing shallow depression in the land and be pebble-lined to imitate a creek-bed. Plants along the sides of the creek-bed would overhang and soften the edges, and some reeds and sedges could be grown in the creek-bed itself.

We planned to leave in its grassy state the south-eastern quadrant where the water flows in. We were aware of the need to do nothing that would impede the natural drainage flow

A problem area turned positive, using a little imagination.

and thought our concept would actually improve it, while trapping some of the water in the pond.

The area between the creek-bed and the eastern fence would be planted with native and indigenous shrubs, ground covers and grasses. There would be nothing too tall here that could eventually grow to block our view. The existing gum trees would provide the height dimension, and the understorey

The first plants in what will be a native shrubbery along the eastern fence.

we intended to plant would be a wonderful bird habitat.

If your block has a similar challenging area, open your mind to the possibilities it gives you to create a stunning feature. Here is where your book and magazine browsing and garden visiting can really pay off.

Vegie plot

We had grown most of our own vegetables and fruit for 30 years and it was one of our major goals to continue to do so, even if on a smaller scale than previously. The area bounded by the north and west fences, the house to the south and the shed to the east seemed ideal. Sunshine for most of the day and shelter from all winds made it quite a separate microclimate within the garden.

Beds could be built up along the corner section of north and west fences, and a roughly central bed between the fences, the house and the shed. The fences would support some kind of mesh for growing peas and beans. In the middle of the central bed I would have a four-grafted citrus tree on dwarfing rootstock so it would not grow too large.

In the same physical space, but separated visually by a corner of the house, I had a perfect north-facing, sheltered location for some fruit trees. The plan was to put lattice on the end of the deck on which to espalier two four-grafted stone fruit trees. In this way I could control their growth and easily keep off avian fruit thieves.

An added bonus of this area was that I still had a sheltered, north-facing brick wall to

This area beside the driveway soon became a shrubby copse of hardy plants attracting insects and birds.

provide a good microclimate for those favourite tropical hibiscus plants.

If growing fruit and vegies is not one of your priorities, a similar spot in your new garden could become a children's play area, a private sunny retreat or an entertainment area. It could even be a good spot to site that swimming pool, if you are still determined to have it and water restrictions in your area make it feasible.

Driveway plots

The driveway is semi-circular. On the western side there is only a narrow strip between fence and driveway. I had no specific ideas for this patch, but knew some nursery visits would be needed to seek out suitably hardy plants that would not grow too large and cause access problems in years to come.

On the eastern side, between driveway and grassy water-inflow area there was enough space for a shrubby copse. I envisaged native and similarly hardy bird and insect-attracting species here. Right beside the driveway, a section had been bordered with cut chunks of a dead red gum we inherited with the block. Behind and between the chunks a bed could be built up for some heat-tolerant species.

This area is in dappled shade from the existing trees for most of the morning and again in late afternoon, but the bed right beside the driveway would be subject to radiant heat from the stone surface, so was no place for hydrangeas.

Careful thought needs to be given to selection of species for bordering driveways or paths. Any plant with a naturally bushy or weeping growth habit will need constant pruning to prevent access problems. Prickly or thorny plants might be unsuitable for safety reasons.

Plot between house and creek-bed/pond

Though what is conventionally considered to be the front of the house faces south onto the road, we consider the east-facing side to be our 'front' as it has large windows overlooking a hill view and gives access to the deck, our outdoor sitting and eating spot.

The eastern and northern aspects of a garden are often good places to consider planting deciduous trees that will shade the house in summer, but allow in the winter sunshine. In our case, the view is to the east and we placed a higher priority on maintaining it than on planting tall shade trees that would block it out – not a practical decision, I know, but practicality shouldn't always win out over aesthetics. In fact, we were determined to have both, so began searching for a couple of small trees that would shade the deck without obscuring the view too much.

Here we initially intended to plant low shrubs (mainly natives), ground covers, herbs and irises – a long-time favourite – with informal paths meandering through. A section of paving running along the house and leading to the shed and the vegie garden access gate was planned for the future. The planting design would have to be taken from the line of the path. In the meantime, we simply left the grasses and weeds that existed when we bought the place and kept the area mown for neatness.

Palm plot

Family holidays in the tropics and subtropics had inspired a love of palm trees. We were

Early palms which were killed by frost and had to be replaced.

determined to find somewhere in the new garden to incorporate a few specimens. A bed along the front of the deck with a blaze of colourful annuals growing around the bases of a few palms seemed like a good idea. We would be able to admire them from the house and when we sat outside, and we hoped they would frame the view rather than interfere with it.

Shed plots

Our shed was built before the house so we would have somewhere secure to store materials and tools during house building. Before deciding where it would go, we asked our architect to draw it in so we could see it in relation to the house and fences.

It wasn't until we had lived in the house for several months that I realised the planting possibilities of the shed surrounds. Not only would garden beds along the east- and south-facing shed walls soften them and help blend the shed in with the garden, it was the nearest I had to a shady spot. Dappled shade from one of the gum trees in the morning and the shadow of the shed itself in the afternoon made what I thought would be a reasonable area in which to plant some shade-loving plants I'd brought from our old garden.

A plot summary

We bought the land and had the house built during what was the worst drought in 20 years, or living memory, or a hundred years, depending on where you live. By the time we moved into the new house in December 2002, it was the tail end of a very dry time, with severe water restrictions. Though this meant we had to delay establishing the garden, it also gave us some thinking time to clarify ideas and find suppliers. The garden plot concepts described above emerged during this time. Some were fully formed. Others needed more work but were well on the way.

I realised that I really didn't need to draw up a plan, as is usually recommended, but could work very easily from my mental images, backed up by reading, other research and previous experience. When it finally rained and garden creation could commence, I started with the plots that were either very easy to do or had the most fully formed concept. Chapter 9 will detail the development of the plots, how I created each one and an appraisal of the results.

Future maintenance

Before you become too serious about planning your new garden it is wise to make a realistic appraisal of how much maintenance you are likely to have time or inclination for in the future. If your busy life leaves little gardening time, there are some specific features you need to think carefully about.

Hedges

Formal hedges need constant attention to keep them in shape and growing evenly. If this sounds like a problem in the making to you, either ensure your plan has no hedges or choose a more informal loose hedge. Lavender and rosemary are great multi-use herbs that will give you fragrance and flowers. New Zealand flax, dianella or dietes make a strappy hedge, and New Zealand flax comes in a wonderful array of leaf colours for an all-year display. Diosma, thryptomene and astartea will respond to occasional haircuts with the hedging shears to give a fine-leafed hedge with delicate flowers. Grevilleas and dwarf wattles will provide food and habitat for nectar-loving birds and insects. For a low-growing, loosely manicured hedge with glossy foliage and bright pink flowers try *Escallonia* 'Newport's Dwarf'.

Roses

Roses are perennial favourites, but are very high maintenance and look dead and ugly for several months of the year. Do you like their flowers enough to make the work seem worthwhile? If the answer is no, there are many attractive shrubs that need far less maintenance.

Ponds

Ponds require a lot of work and expensive additives to keep them looking crystal-clear. Does your situation lend itself to a natural clay-lined seasonal pond in which the soil and water biota do all the work for you? If not, maybe that spot designated for a pond would be a better place for a gazebo with a small water feature incorporated.

Pots

Plants in pots need more care, especially watering, than plants in the garden. If you love pots, keep them big, use the very best potting mix with water-saving granules incorporated, group them in clusters and try planting a variety of succulents in them for a stunning look that needs little care.

Hard surfaces

Try to keep hard surfaces to a minimum. Gravel and pebbly materials quickly become scattered where they are not wanted, especially when there are children in the home, and often begin to look unsightly when weeds start to grow through. Concrete, bricks and pavers need constant sweeping and maybe even frequent washing, depending on the situation. An electric blower/vacuum device can make maintenance of hard surfaces less onerous.

Lawns

The lawn is often the highest maintenance area of a garden. Avoid planting one altogether, or consider some of the alternatives discussed in Chapter 2.

Drawing up a plan

Some people really need to get ideas down on paper before they are confident they are doing the right thing. If this includes you, or if you have a very large area to work with or you envisage a complicated design, try the following. Even if you prefer to work from mental pictures, these ideas will help clarify your thoughts. Read right through to the section on design details before putting pencil to paper.

- Buy a large sketchpad. You can use graph paper for greater accuracy, but you're not designing a high-rise tower and your future plants won't know they are meant to keep to four squares of graph paper, so don't worry about being too precise. Draw an outline of your block and mark in, roughly to scale, house, shed, garage, existing paths, fences or trees. You might have access to the builder's site plan you can photocopy. Use arrows to show the compass points.

- For ease of changing, mark in any further features you would like to add in pencil – outdoor living area, paths and paved areas, water feature, pond, play area, shed, pergola. Remember too, the essential but unattractive clothesline. You can make cardboard cut-outs of these and move them around on the plan until you are satisfied.

- Ideally, for economy and convenience, any project that needs machinery access (paving, for example) should be done at the appropriate stage of house building. If this is not possible, make sure you locate these areas on your plan in an accessible position.

- If you have existing outdoor furniture, ensure that your proposed entertainment area is large enough to fit it and to move around comfortably. Similarly, if you plan to incorporate a children's play area, allow space for equipment to be used safely. A swing, for example, does not take up a large amount of space in itself, but needs a good clear surrounding area to be used without mishap.

- Once you have decided on placement of the hard areas and infrastructure, make a more permanent copy of the layout. In fact, make several to play around with.

- Mark garden beds on this copy. Remember that sweeping lines not only look more pleasing to most people and give a feeling of spaciousness, they are easier to mow around than tight curves or angles. Play with different lines until you find a design that pleases you – this is why you made multiple copies.

- You should know by now any areas of the garden that are shaded at particular times of day. Write this in to help with plant selection.
- Plan on grouping plants with similar water, sun, shade, soil and nutrient needs to simplify future maintenance.
- As you decide on plant species they can be marked in. Alternatively, mark the shapes of plants you would like for each plot and use this as a guide when you visit nurseries searching for suitable species. Start with a pencilled outline and add colour later to indicate flower or feature foliage colours. This helps you work out a harmonious, or contrasting, colour scheme, depending on your preference.
- If any of your proposed garden beds has sandy, clay, acid or alkaline soil that will need attention before planting, key this in also.
- There are computer programs for garden design. If this appeals to you, these hints can be adapted to computer use.

Design details

Don't be daunted by the thought of designing your garden. This is where you get to have some fun and express your personality. By this stage of planning, you will already have incorporated many design ideas, probably without realising it. No one is grading you on this project; you can please yourself. These simple suggestions might help your creativity flow.

- If you have decided on a particular style or combination of styles, do some reading to choose plants, colours, furnishings and accessories to complement the theme(s).
- Repetition of shapes, colours, lines and materials can unify an area, or the whole garden, but too much sameness can become boring and is not good for biodiversity.
- In a formal garden create symmetry by, for example, having mirror-image beds on each side of a path, or by having a central feature with symmetrical beds radiating from it.
- A feeling of balance in a less formal garden can be created by clever placement of beds of similar shape, mass, planting design or colour scheme.
- In a new garden the height aspect of design is lacking and can take some time to fully develop. This can be overcome to some extent by incorporating a few advanced plants, an arch or an obelisk or two.
- Where space is tight, light colours, a path wandering into the (pretend) distance, a *trompe l'oeil*, or a feature placed to attract the eye to a distant point can all make an area appear more spacious. Another way of creating a spacious feeling is to plan your garden so a distant view, or a neighbour's attractive garden, can be seen.
- To close in an area, either for privacy or to create a cosy garden room, use hedges, vine-covered trellis, an arbour, or grow vines on existing fences.
- Textural variety can be added by using plants with dramatic architectural leaves, large furry leaves, shiny or tiny leaves, or by varying the texture of hard surfaces and ornaments.

- Use paths to draw the eye to a feature as well as to encourage easy access and exploration around the garden. Paths need not be hard surfaces; they can be grass, pine bark, sawdust or other soft (and inexpensive) material.
- Remember that foliage colour can be as dramatic as flowers in providing seasonal variety and interest.
- Flower and foliage colour are often quite short-lived, so choose plants to give colour, and feed the beneficial birds and insects, year-round.
- Simple flowing lines create more visual impact and are easier to maintain than lots of tight curves and angles. However, if curves and angles are your preference, have fun with them and exaggerate them to make a statement.
- Incorporate surprises: a statue around a corner or peeping out from under a plant, a pot plant as part of the edging, an unexpected vista opening up when deciduous trees lose their leaves.
- Try not to use too many hard impervious surfaces for paths, driveways and outdoor living areas. They prevent the soil from absorbing water and contribute to localised flooding during heavy rain. In addition, the more hard surfaces you have the less space there is for real gardens. Consider too, that concrete, bricks, tiles and pavers will need to be swept regularly or otherwise kept clean, so if time for maintenance is an issue with you, minimise use of these surfaces.
- The entrance to your garden welcomes you home and invites visitors in. Make it appealing by incorporating special features: large potted plants, sculptures, an arch or pergola, an antique or handmade gate or a mosaic path are all possibilities.
- Use water to encourage biodiversity: still water for tranquillity, rippling water for relaxation or splashing water for liveliness.
- If planning a lawn area of whatever sort, ensure that plants and decorative elements will not unduly restrict mower access.

How much DIY?

This is a good stage to consider how much of the work you can do yourself and how much requires hiring a professional. Most garden jobs are within the capabilities of an average handyman, but they will take longer and the end result often will lack the professional standard of finish. If your garden style is casual and your budget is tight, this should not be of concern. If you prefer a more formal style and have a more generous budget, professionals are worth paying the cost.

Acknowledge your own physical limitations. It is surprising what you can accomplish with a

Some jobs are just too big to tackle yourself.

trolley, a wheelbarrow and a strong lever (like a crowbar), but some jobs are simply too heavy or too difficult. It is not worth breaking your back in order to save money. Either revise your plans to a smaller, simpler scale that you can cope with, organise a working bee with a group of your strongest friends, or set a small amount of cash aside each month until you can afford to pay professionals. Meanwhile, work on those areas you can do yourself.

Above all, be realistic. Don't expect to achieve the standard and speed displayed on television programs. In the real world gardens take time to achieve and grow slowly.

Observation

Don't be in too much of a hurry to rush out and buy plants. Live in your house for a while first. There will be enough to do getting settled in your new location. Time spent now in observation, planning and thinking will prevent costly mistakes and ensure you are happy with the results.

Observe the weather patterns, areas of sunshine and shade, where water lies after rain, what grows well in similar positions in neighbouring gardens. Note too any windy spots and the direction of the wind. Decide if and how you can modify strong winds. If this planning period occurs in winter, notice where any frosts form and note on your plan that you will not plant frost-sensitive species there.

You still haven't bought any plants.

Start making compost as soon as you can.

MAKE COMPOST

As soon as you have moved into your house, buy a compost bin and start composting kitchen scraps. Make sure you get one that has plenty of air holes – good aeration ensures quicker compost and fewer odours. With only kitchen scraps going into the bin, it will take months to fill. Every few weeks sprinkle in some manure, blood and bone or Dynamic Lifter, and a layer of dry matter such as straw or sugar cane mulch. Keep the contents damp, but not soggy. With a well-aerated bin, this might necessitate watering the contents a couple of times a week during hot weather.

Once the bin is about half full, use a garden stake or crowbar from time to time to drive holes down as far as you can into the decomposing material to allow more air in. Do this every couple of weeks. When the bin is full, leave the contents to decompose and buy another bin to start the process again.

Once your garden is going and you have garden refuse to add as well, the bin will fill more quickly and you will have an ongoing supply of free, organic fertiliser and mulch teeming with beneficial micro-organisms.

Tiding you over

Perhaps you are living in your new home by now. Congratulations! More likely than not you are becoming impatient to create a thing of beauty from your bare surroundings. This is usually a time when both money for major projects and time to undertake them are in short supply. You might not have your plan ready yet. Even if you do, there are still some steps to take before the creative fun can begin. You are probably not going to want to spend much time or money on temporary beautification works, but there are some simple things you can do to make your garden a more pleasant place.

- If piles of leftover sand and builders' rubble are an eyesore, make it a top priority to get rid of them and at least have a clean, smooth surface. Unless you own heavy earth-moving equipment, this is a job you will have to pay someone to do. Look for contractors with a backhoe and tray truck. Spread any sand with a rake or set it aside to work into any areas of heavy clay.
- If you have the opportunity, set aside any leftover building materials that could be useful later. Bricks or flexible lengths of timber make good edgings, and a brick path that matches your house is a definite plus. Screenings might be useful for drainage. Steel reinforcing mesh can become a climbing frame for peas or beans.
- In the majority of cases, the mess will have been cleaned up before you take possession. You could just have a jungle of grass and weeds. Keeping it mown can make a big difference. You'll be surprised how green and presentable even the weediest patch can look after a few weeks of regular mowing.
- Maybe you have potted plants brought from your previous garden. Place them where they will make an impact – along a path or fence, on a deck, where you can see them from the house.
- Strategically placed outdoor furniture, with an umbrella and a colourful pot plant on the table will make an impact and give you somewhere to relax. If you don't already own an outdoor setting, this might be a high priority and can be moved around the garden as future developments are underway until it finds a more or less permanent home.
- A few cubic metres of aesthetically pleasing mulch spread around will not cost much and might transform an area. Do put sheets of cardboard, newspaper or weed mat under it to prevent weeds growing through. When the time comes, the mulch can be raked into a more permanent location, so the cost will not have been wasted. Even if it has decomposed, it will have added organic matter to the soil and improved its structure.

ACTION PLAN

- Determine your garden style.
- Decide where plots are naturally defined.
- Draw up a plan or clarify your mental pictures.
- Play with design ideas.
- Decide what you can do yourself and what you need to pay professionals to do.
- Try some temporary beautification tricks.

- Similarly, use a long-lasting mulch such as wood chips to form temporary paths.
- All right – *now you can buy some plants*, but only a few. Seedlings of colourful annuals are cheap, look cheerful and can be kept in pots in strategic spots. As you decide on plants you want for permanent beds, buy them and keep them in their pots temporarily, but only after reading the next chapter on plant selection.

OBTAINING YOUR PLANTS

With your garden plan well on the way, it is time to do some serious searching. In some cases you will have a definite list of plants for particular plots. You might know what sort of plants you want – size, shape, colour and degree of hardiness – but not have specific species in mind. Other times you might visit a nursery with just the shape and location of the plot in mind, or on paper, hoping to find inspiration.

Economical gardening

You don't need to spend a fortune. There are many sources of free or cheap plants.

- Friends, relatives and neighbours are usually only too willing to give you cuttings or divide plants for you. In many cases these offerings will not be the plants you had in mind, but accept them graciously anyway. They might turn out to be more suitable than you first thought and at least will provide some greenery and flowers while you are in the development stages.
- Small plants are usually much better value than expensive large ones. They cost much less, will adapt better to transplanting, are easier to handle and will grow quickly. If you are really determined to make a quick

impact, just a handful of large plants in a strategic area are all you need.
- Many annuals are available in punnets. These are very economical, great for quick splashes of colour and need not be a permanent feature.
- Seeds are even cheaper. Scatter some in a weed-free bed or large pot and keep the soil moist. You'll end up with dozens of plants to spread around.
- If you are moving from an established garden, make time in those frantic last few weeks before moving to dig up plants, separate root divisions, take cuttings and collect seeds. This will give you a head start in the new garden. If you are not going to be in a position to care for multitudes of pot plants for a while, give them to friends on the understanding that you will be able to collect propagating material when you are ready.
- Shop around. The most obvious source is not necessarily the cheapest. When the time came to buy our palm trees, for example, we decided on the species and rang some specialist palm nurseries for prices, only to find that not only was our first choice unavailable indefinitely, but the price seemed very high. We thought we were clever when we bought our second choice of species from a large hardware

chain at about a third of the nursery price. A few weeks later we saw the same plants slightly cheaper again at a local monthly market. This market has proven to be an excellent source of reasonably priced plants.

- When buying clumping or matting plants, look for any that are advanced enough to be readily divided to give you several. In this case the more advanced plant is better value than the smaller one. Some advanced plants that might easily be separated to give you several new ones include: irises, iris japonica, native grasses, phormiums, reeds and sedges and numerous herbs. If you plan on growing succulents, specimens often have 'pups' growing from them that can be easily separated.

Finding that special something

You might have a specific species or variety in mind that is not available from local sources. Read the ads in gardening magazines; don't just skip over them to admire the pictures. They are a very useful resource. Contact any advertisers who might stock the plants you are seeking and ask for their catalogues.

Plant catalogues are a wonderful source of information, often telling you plant size and climatic and watering needs as well as prices. Most mail-order plants are quite reasonably priced, but, if they seem expensive to you, make a list of those you are interested in and take it with you on your gardening expeditions; perhaps you can obtain them cheaper elsewhere. If a thorough search of other likely sources proves unsuccessful, you need to choose whether to pay the price or substitute a more accessible and cheaper plant.

The internet is a useful tool when searching for anything you want to buy. Try typing the name of the plant you are after into your search engine.

Indigenous plants

One of our major aims was to encourage biodiversity within our garden. To do this we wanted to obtain as many indigenous plants as we could as an understorey to the red gums, and combine them with other natives and exotics with similar food and habitat potential.

Finding out about the indigenous plants of this area was a challenge. Some progressive councils publish a list or booklet. Our council had no relevant information at all. A local native plant nurseryman stocked only a few indigenous plants. I managed to obtain a list from the local environment group whose members propagate indigenous species, but no information about their habits or specific

Varnish wattles (*Acacia verniciflua*), planted beneath the red gums on the nature strip, are indigenous to our area.

needs was available and plants had to be ordered ahead for the next season.

Finally, success. On a visit to a nearby environment expo I found a comprehensive tome published by what is now called the Department of Sustainability and Environment (DSE) containing everything I needed to know and more. From this I compiled a list of species that would suit my needs and took it around to nurseries everywhere I went. In this way I was able to obtain at least some indigenous plants from which I can propagate in years to come.

The moral of this story is, when searching for indigenous plants and relevant information, if all else fails, try a government department – you could be pleasantly surprised.

'Orange Twists' – one of our major successes for a difficult area.

Accidents and inspiration

Though your garden plan might by this time be quite well developed, leave room for accidents and inspiration. A nursery visit could very well introduce you to a new variety you hadn't considered, or a gift from a friend start a new line of thought altogether.

Very soon after we moved into our new home – remember this was the tail end of a major drought – a friend who knew I intended to plant succulents in one area was digging some overgrown specimens out of her garden. Though the soil here was rock hard and I was not ready to plant anything, when she offered them to me what could I do but accept gratefully. I scratched out a bed as best I could and placed them in a temporary spot in the future succulent plot. They coped well with the harsh conditions and looked so good once some other development took place around them that they are still there.

It was a while before any concrete ideas began to take shape for the narrow plot on the western side of the driveway. We visited several nurseries in search of inspiration. Finally, we found just the right plant – 'Orange Twist', a compact form of *Syzygium australe* (brush cherry, related to lilly pilly). It is column-shaped, growing to three metres high and spreading two metres. Its new foliage is orange coloured and slightly twisted then fades to a glossy deep green. It has small white flowers in spring to attract insects and bright magenta berries in autumn to feed the birds. We can even make jam from the berries if we want to. The more frequently it is trimmed the bushier it will grow and the more often there will be new orange leaves. We bought six.

Waterwise plants

Water conservation is a major issue for all of us. A little forethought and planning can

Succulents are a great solution for a hot-dry zone.

ensure that our gardens are water-efficient from the start, and this begins with the type of plants we choose. Most indigenous species (apart from those adapted to damp shady conditions) will thrive on a low-water regime once established. They will need regular watering for the first year or two. After this, summer watering can be reduced gradually and watering for the remainder of the year almost eliminated.

Cushion bush (*Leucophyta brownii*) with its scaly silver leaves is an outstanding example of a water-efficient plant.

When you go to a nursery to select plants you will find that some labels have a reasonable amount of information about growing conditions, or at least a key to water needs, but this is not always the case. There are some clues you can look for, mainly in the type of leaves plants have, that indicate tolerance of dry conditions.

- Cacti and succulents store water in their swollen stems and leaves.
- Grey and silver-leaved plants reflect sunlight, as do light-green, blue-green or grey-green leaves to a lesser extent.
- Small, tough, round or needle-like leaves have minimum surface area exposed to the sun.
- Leaves have small pore-like openings, called stomata, through which water vapour and other gases pass. Hairs on leaves slow wind movement over the stomata and reduce moisture loss.
- Tough waxy leaves have reduced water loss through the leaf surface.
- Narrow upward pointing or drooping leaves have less surface area exposed to the sun. Some leaves even have the ability to change direction during the day in order to maintain minimum sun exposure.
- Plants with hard woody branches and stems are less inclined to wilt and show heat stress than those with a tender stem structure.
- Water-efficient plants have leaves with few stomata or stomata on the underside of the leaves, thus lose less water during periods of heat. Stomata are so tiny you are unlikely to see them unless you take a magnifying glass to the nursery. Plants with few stomata include eucalypts, wattles and banksias.

Plants to avoid

As well as environmental weeds, mentioned in Chapter 2, there are some other things to be cautious about when choosing plants.

Poisonous plants

Over aeons of evolution many plants have devised ingenious methods to prevent them from becoming the next meal for a variety of herbivores. These survival strategies include poisonous seeds, berries, leaves or sap. Some plants are poisonous in all parts. Other plants, while not being harmful to most people, can cause allergic reactions in those with sensitivities. If there are young children in the household or who frequently visit, either don't plant these species or supervise children while they are in the garden. The following are some common ornamentals to be cautious of, but there are many others. The best strategy could be to train children from a very young age to recognise what they *can* eat – herbs, vegetables and fruit – and leave everything else alone.

- Angel's trumpets – all parts poisonous
- Arum lily – all parts
- Autumn crocus – leaves, seeds and bulbs
- Azaleas – leaves, flowers and pollen
- Belladonna lily – sap from bulbs and leaves
- Box tree – stems and leaves
- Castor oil plant – seeds, even from touching
- Columbine – seeds
- Cycads – berries
- Daphne – all parts
- Diffenbachia – sap
- Euphorbia species – sap
- Hellebore – sap
- Hydrangea – sap in prunings
- Japanese windflower – all parts
- Lily of the valley – berries
- Oleander – sap and leaves
- Plumbago – sap
- Privet – all parts
- Rowan – seeds
- Spindle tree – all parts
- Viburnum – berries
- Wisteria – pods and seeds.

In cases where the sap or all parts are poisonous, it is wise for even adults to handle these plants with care and wearing gloves.

Prickly plants

Many plants have prickles as defence mechanisms and some native prickly plants make ideal bird habitat because birds nesting or roosting in them are safe from a range of predators. When planting any prickly or thorny species, even roses, make sure they are well away from children's play areas. Roses are frequently planted along pathways, but for the safety and comfort of family and friends, make sure they are not too close.

Poolside plants

Avoid planting anything close to a pool that will drop berries or leaves into the water. This is only dangerous if the dropped parts are poisonous (as above), but will be a continuing nuisance at particular times of the year. Fruit and berries might stain the pool coping and surrounding paving as well as being slippery to walk on. Don't plant anything spiny or sticky.

Gum trees will continuously drop leaves and twigs. Deciduous trees will drop their mass of leaves at one time of year, which might be preferable to a continuous drop. Trees with invasive roots are another no-no (see below for trees to avoid planting near drains and sewers).

Large plants near the house

It seems too obvious to mention, but I see it so frequently that it is apparent many people just don't think of it. Large plants right next to your home are at best a nuisance, at worst a danger. That tiny elm in a tube will grow to be a giant that will block light from your windows, drop leaves in the gutter and branches on your roof; its roots might even damage the foundations and clog the drains. Read plant labels before you buy. If you have space for big trees, great, but not near the house.

Plants near drains or sewers

Avoid planting any large trees close to drains or sewer lines and only plant shallow-rooted species near any trenches as plant roots will naturally grow into the well-aerated soil back-filling the trench. Keep the following plants at least four metres from sewers or water pipes.

- Bauhinias
- Brachychitons
- Brush box
- Casuarinas
- Coral trees
- Cotoneasters
- Eucalypts
- Figs
- Flowering tamarisk
- Golden ash
- Hakeas
- Liquidambar
- Maples
- Monkey pod tree
- Native frangipani
- Nettle tree
- New Zealand Christmas tree
- Paperbarks
- Persian silk tree
- Pittosporum
- Privet
- Rowan tree
- Strawberry tree

- Wattles
- Willow myrtle
- Willows.

While you are choosing and buying plants, progress on many fronts can be taking place and your garden could get very messy for a while. Relax. This is just a stage on its way to looking very good. In the next chapter you'll find out how to steer your way through the chaos to achieve your goal.

ACTION PLAN

- Source free and cheap plants.
- Track down less-common species.
- Find out about indigenous species.
- Choose waterwise plants.
- Avoid poisonous plants, prickly plants where they might be a hazard, large trees close to the house and trees that could interfere with drains.

PREPARATION AND CONSTRUCTION

All your observation, research and planning now can be put into action. Before going ahead with any above-ground works there are some points you need to consider, including some important aspects that will never be seen in the finished garden.

Locate existing underground services

Before you begin moving earth around you need to know where stormwater drains, water pipes, inspection pits, septic and sullage tanks, sewage lines, gas lines or underground electrical cables are located. If your builder, plumber or electrician cannot supply this information, or you are not in a position to even know who they are, approach the relevant authorities: council, water board or electricity supply company. This is not relevant if you are going to be planting into raised beds.

Remove unwanted items

If there are existing plants, you must decide whether you want to keep them where they are, move them to another spot or remove them entirely. There might be dead tree stumps, collapsing fences or sheds, broken old paving or building detritus to be removed. Make these decisions and carry out removals before commencing new work. It might be possible to salvage old materials to incorporate in the new garden. We inherited some chunks of a fallen red gum trunk. We built a garden bed around one and used others to line a section of driveway. We kept the property's original rusty farm gate, stripped the wire from it, painted the frame to match one of the house colours and used it for a decorative backing to a succulent bed. A rusty wheel from a piece of agricultural equipment I found when digging in a vegie bed now rests against the gate frame. To me, these little reminders of the past add a sense of history to the garden. In addition, we used leftover building materials as edging for a number of beds.

Install drainage

In the vast majority of cases excess water will be carried away by natural runoff and seepage into the soil. You can help this process by ensuring your soil is well aerated and that garden beds and lawns slope, where possible, away from structures and towards gutters. However, there are several reasons poor drainage might occur in some situations and each will need a slightly different approach. If extra drainage is necessary, it must be done before anything else.

High water table

A serious drainage problem such as might be caused by a high water table that floods your garden after every heavy rainfall needs professional attention to lay underground trench drains or perhaps an absorption pit to remove the water. Permission from your local water authority might be needed.

Periodic flooding

Periodic flooding of an area – this occurs on our block because of inadequate town drainage – can be dealt with in two ways. One way is to obtain the relevant permissions from the local council and water authority and have drainage installed to their specifications. The other option is to make a feature of the

After heavy rain, a section of our block floods. We made this into a feature.

'problem' by creating a pond, ephemeral creek or bog garden.

Waterlogged clay

There are a couple of solutions for areas of heavy clay soil that become seasonally waterlogged. You can either dig sand and/or organic matter into the topsoil, or build raised beds that will shed water and allow plant roots to remain aerated. Depending on the severity of the problem, a raised bed might need to have drainage pipes laid underneath it or have a generous thickness of coarse screenings at its base. If in doubt, obtain professional advice.

Claypan

There might be a hard claypan not far beneath the surface that will slow water drainage and contribute to surface waterlogging as well as prevent plant roots from penetrating to deeper levels of the soil. You might choose to deal with this the hard way by breaking up the claypan with a mattock and crowbar, or by hiring a rotary hoe to do the job.

I encountered this situation in an area where I had made a raised bed of imported topsoil. If I had simply planted into the topsoil, as I had first intended, the roots of my young plants would probably have been unable to penetrate the claypan. I scraped a hole for each plant in the new topsoil and filled each hole with water. I did this twice to soften the hard earth beneath. Then, with water still lying in each hole, I used a crowbar to break up the claypan. The sound of the water being slurped up by the thirsty soil indicated that the impermeable layer had been broken. I was then able to work in some compost and decomposed manure and proceed with planting.

Install new services

Your plans might call for plumbing or electrical work in some areas, maybe more taps, pumps, irrigation pipes or lighting. Lay pipes or cables for these underground before paving or garden beds go in. The ready availability of solar-powered lights and pumps means this is not as important a consideration as it used to be, but you might still need to lay irrigation pipes underneath a path or lawn, or connect electricity to a shed or workshop.

Allow access for future construction

You might need to get machinery and equipment into the back of the block, perhaps to construct a shed, to dig a pond or to lay paving. If this is the case, it is unwise to plant anything at the front that will probably be damaged or need to be replaced later. Do the construction work first. If you can't afford to pay for major works at this stage, use some of the temporary measures described in Chapter 3.

Arrange a dumping spot for materials

It is often necessary to take delivery of loads of soil, sand, screenings, pebbles, rocks, mulch or building materials. Arrange a convenient spot for these to be dumped. The neighbours will probably not appreciate your garden-in-the-making cluttering up the

Our works supervisor made sure materials were dumped in a suitable spot.

nature strip or footpath. Unless you are home to direct them, delivery drivers will offload materials in the most convenient spot for them, which might not suit you, especially if you then need to move it all out of the way.

Some jobs will take a while to complete. If a load of pebbles, for example, is dumped on your future vegie garden area, as was the case with us, no work can be done on that area until the load is moved. Progress in one area is thus often dependent upon completion of another.

Buy some tools

Novice gardeners will need a few tools to begin with. There is no need to buy a separate piece of equipment for every job you can think of; a few basic items will see you through. Don't choose the cheapest tools available, you'll probably find them to be a false economy. On the other hand, for the amount of work needed in most domestic gardens, top-of-the-range equipment is not needed. Here is my list of gardening essentials:

- spade
- metal rake
- mattock
- crowbar (not needed if you have sandy or loamy soil)
- trowel
- wheelbarrow
- mower if you have lawn
- garden fork if you need to aerate compacted soil
- trolley for moving heavy loads, even bags of mulch or compost, so you don't strain your back.

In time you will need to buy secateurs, hedge shears and other pruning equipment and some form of edge trimmer.

Unless you intend to lay paving and undertake construction work yourself, these few items will cope with most jobs. More specialised tools and equipment that you'll only have limited use for can be borrowed or hired at the appropriate time.

Order of work

Now it is time to plan an order of work. You might decide to start with an area you think requires the least work, or one that will make a big impact, or one that is clearly planned. Maybe you are the sort of person who would rather get the biggest and hardest job out of the way first. In any case, as discussed above, where extra drainage is needed this must be done first. While the ground is chopped up, it makes sense to install any underground pipes or cables at the same time.

Construction and hard surfaces

I cannot stress this enough: if you can possibly afford to have sheds, pergolas, ponds and pools constructed or paving laid before making and planting garden beds, do so. It is far easier to get the mess, piles of materials and builders' equipment out of the way before planting. The workers will have a clear field to work in and you won't be worried about plants being damaged in the process. In addition, it is usually the case that lines of garden beds flow from or are framed by the structures and paving and it is very hard to work from imaginary lines. Even lines drawn on the ground or marked out in string are not the same as the real three-dimensional structure.

If your handyman skills are basic or nonexistent, pay professionals to do construction and paving jobs. If your skills are up to the job, you can save a heap of money by doing it yourself. There are many detailed 'how-to' books available for all types of garden projects. The following hints will be especially useful for anyone who has not created a garden before.

- Having the right tools for the job is half the secret to a professional looking finish. Ask advice from sales staff if you need to buy tools and are not sure of the right ones. You can recoup much of the cost of expensive tools that you don't expect to use again by selling them afterwards. This could work out cheaper than hiring tools in some instances – do your sums.
- When paving, start with a flat, firm surface that slopes slightly away from

A simple string level can be used to ensure a level foundation. This concrete footing had a brick edging laid on top.

buildings. You need a base of brickies' sand and a compactor to make sure it is firmly tamped down.

- Ensure levels are right, lines are straight and the foundations of any job are square before proceeding.
- Work out quantities of materials carefully, double-check, make a list for each job and ensure everything you need is on hand before starting work.
- Compare prices before buying; you can save by shopping around.
- When employing professionals, get quotes, be quite clear of what the quote includes and hold some money back until you are satisfied with the standard of completion.
- You can save some money, if hiring professionals, by labouring for them yourself and by making sure all materials are ready so no time is wasted. Maybe you will learn enough from the experience to feel confident about tackling the next job on your own.

Garden beds

Don't skimp when it comes to preparing the soil for planting. Poor soil preparation is one of the most likely causes of disappointing plant growth. Look back to the discussion about soil in Chapter 1. Now is the time to put that knowledge to use. Remember:

- Raise pH of acid soil by adding dolomite, lime or wood ash.
- Lower pH of alkaline soil by adding sulphur or iron sulphate.
- Add gypsum or one of the new liquid clay-breaking products to heavy clay soil.
- Work clay into very sandy soil.
- Add lots of organic matter to both sandy and clay soils.

With these points in mind, there are several ways of preparing garden beds: digging, sheet mulching or building raised no-dig beds. Whichever method you use, begin by marking out the proposed area with a garden hose, string or marking spray paint.

Digging

This is my least preferred option and I avoid it as much as possible. It disturbs the soil biota

Imported topsoil in the raised vegie bed was well mixed with fertilisers and organic material.

Applying gypsum to our no-dig beds.

and diminishes its beneficial effects, can damage soil structure and is simply hard work. On the other hand, if done once only, it is a good way of loosening and removing weeds without using herbicides and of aerating compacted soil. It also enables you to incorporate any additives that the soil needs by digging them in. Hard as it has been for me to accept, I know there are people who actually enjoy digging. I urge these stalwarts to remember that more is not necessarily better; repeated digging will deprive your soil of the benefits of myriad valuable soil biota and can cause a hard pan to form below the level normally dug to.

Once the bed is dug over and the weeds removed, rake the surface level. Scatter over any necessary additives and rake the surface again to distribute them evenly. Water well. Your chosen watering system can be installed either now or after planting (see Chapter 7).

Sheet mulching

This is an easy method of having a garden bed without really making a garden bed in the conventional sense.

Initial hard work constructing no-dig beds does pay off!

- Mark out the area.
- Closely cut grass and weeds.
- Scatter over any necessary additives and rake evenly.
- Water well.
- Cover it all over with a thick layer of newspaper, cardboard (television and refrigerator boxes are excellent) or one of the purpose-manufactured weed mats. Wet this layer.
- Cover this layer with a cosmetic layer of pea straw or other organic mulch. Thoroughly water so the mulch is soaked through and keep the area damp until you are ready to plant. The weeds will decompose under the sheet mulch and the soil biota will incorporate them, and your scattered additives, into the soil.
- Irrigation lines can be laid under the weed mat or on top of it and then covered by the cosmetic layer.

When you are ready to plant into the 'bed', simply make a crosswise cut in the sheet mulch, fold the ends under and dig into the soil beneath with a trowel, removing any weeds from the spot as you do so. Any extra organic fertiliser can be well mixed into the hole at this stage.

Existing areas of trees and shrubs can be sheet mulched as well; you just need to work around them, leaving sufficient space around each plant free of the base weed mat to allow access for watering.

Raised beds

Raised no-dig beds are a dream to work with. There is, of course, no digging involved and the resultant garden bed is friable, fertile, weed free and extremely water efficient.

This photo sequence shows a combination of sheet mulching and raised bed: vegie garden bed brush-cut and edged; thick layers of newspaper are laid; topsoil is added; bed is planted and mulched; close-up of a section of the bed (the peg in the centre is where a four-grafted citrus tree is to be planted).

Raised bed preparation: materials assembled; area of bed now brush-cut and marked out with string; newspaper laid down and dampened; pea straw bales ready to be broken up and spread; first layer of straw is spread and compost and manure raked loosely in; more straw spread on top of other materials; plants go into pockets of topsoil in the straw; the no-dig bed a few months later.

- Start as for sheet mulching by marking out your area and mowing or brush-cutting. If necessary, install drainage beneath or around the bed.
- Sprinkle any necessary soil amendments such as wood ash, lime or gypsum. A scattering of organic fertiliser will help the breakdown of cardboard or newspaper used as weed mat. Rake any additives over the area and water in well.
- Cover the area with weed mat of some sort, either freebies you have collected or a commercial product. This is best done on a wind-free day. If using cardboard or newspaper, overlap it well to prevent weeds growing through and thoroughly wet it as you go.
- If there is edging to go around the bed, this is a good time to lay it, tucking the weed mat securely under it.
- On top of this layer spread thick layers of a variety of organic materials. Some gardeners recommend specific materials, but any weed-free organic materials you have access to are suitable, including pea straw, lucerne in any form, decomposed manures, wheaten or oaten straw, Dynamic Lifter, stable or poultry pen litter, rice hulls, fallen leaves and leaf mould, grass clippings and compost. Remember, a variety of materials will result in a balanced range of nutrients being available to your plants. Coarse materials such as palm fronds, twiggy garden prunings or straw in the bottom layer assist drainage. The layers can be built up twice as high as the edging because they will reduce in height as the material decomposes.
- Water each layer well.
- Alternatively, you can spread bought topsoil or mounds of crushed sandstone over the weed mat layer; the latter is especially good for planting hardy native species.
- Keep your raised bed damp and leave it to decompose for a month or so before planting into it; or, if you are as keen to get plants in the 'ground' as I am, plant into pockets of topsoil straight away.

Edging

Once you have the soil in a garden bed prepared, it is optional to edge it with any of a variety of materials including bricks, rocks, paving tiles, railway sleepers or other timbers, or one of the purpose-made plastic or stone-like edgings that are widely available. Choose an edging that suits the style of your garden or matches your house. Leftover building materials can be put to good use here.

Edging separates garden beds from lawns or adjacent beds, outlines a bed and emphasises its discrete design, keeps organic mulch,

To lay edging, simply dig a shallow trench around the outside of the bed, slightly wider than the width of the material being used. Nestle the material into the soil and backfill, firming the soil down into the trench as you go. Railway sleepers or similar bulky and heavy edging will probably need some small stakes driven into the soil on the side away from the bed to keep them upright, especially if they are holding back a depth of soil or mulch.

Edging can be installed either now or after a bed has been planted.

Lawns

You will have read my opinion of lawns earlier. If you are determined to plant a lawn, and water restrictions allow it, at least choose a drought-hardy grass variety. The secret of a successful lawn is good preparation. If you choose to water your lawn, you might like to install a pop-up system (see Chapter 7). This must be installed before the lawn is planted.

- Begin with a weed-free expanse of finely textured well-dug soil.
- It might be necessary to hire a rotary hoe to dig over the soil. If doing this by hand, several diggings and rakings could be needed to achieve a fine soil texture.
- Rake up and dispose of all weeds.
- Spread any necessary soil additives such as gypsum, dolomite or fertiliser.
- Rake again to distribute these evenly.
- Install the pop-up watering system now if this is your choice and rake soil smooth and flat after installation.
- Water well.
- Sow your lawn seed.

pebbles or raised beds in place so materials don't travel to areas you don't want them, and can help prevent lawn grasses from encroaching into other areas. If you want areas of the garden to flow into each other, you probably won't need to use edging. Some edgings – rocks and bricks for example – can be fiddly to weed around. Smooth flowing edgings make maintenance easier.

- Rake it evenly over the surface.
- Water again.
- The soil must be kept moist until the seed germinates and then while the tender new grass is establishing. For this reason it makes sense to sow lawns during warm, moist weather. It is simply too much of a chore to get a lawn established during the heat of summer, and takes too much water.
- Turf lawns are said to require less water for establishment, but require similar meticulous soil preparation.

Alternatives

In the areas of my garden deemed to be 'lawn', I have simply left the existing weeds and grasses. There is a diverse mixture that includes capeweed, oxalis, creeping buttercup, dock, cat's ear, clover, bindweed, wireweed, winter grass, phalaris, plantain and an unidentified running grass. Throughout most of the year there is always something that looks green, even if only vaguely green during summer. Kept mown, it looks quite presentable, cost nothing to establish and gets no water other than rain. One neighbour admired my newly cut green sward during spring when it looked particularly vibrant and said, 'But I thought you weren't going to plant a lawn'! Despite our love of the green lawn, we need to develop a new mindset that appreciates the changing colours of the seasons and doesn't consider the soft yellow of summer-dried grass to be an eyesore.

Because of their different growth habits, frequent mowing favours grass over broadleaf weeds, so eventually there will be less dock and capeweed and more grass. I will never

The herb lawn area has been brush-cut and gypsum spread. Newspaper is being laid.

have the greenest lawn in the street, but it will almost certainly be the least watered.

Herb lawn

Another area is turning into a lawn of a different type. A mostly native garden on the east of the house ended up taking a roughly circular shape, with raised garden beds surrounding an enclosed flat area. I decided to turn this into a garden room with the shrubs in the raised beds defining the boundary, a dwarf, double-grafted citrus tree in the centre and a fragrant herb lawn as the floor of the flat area. Here is how it's done.

Pea straw is spread over the paper. Holes will be cut in the paper to plant herbs into the soil beneath.

The herb lawn is spreading and flowering.

NO-MOW LAWN

If you wish to avoid mowing altogether, live in a frost-free area and have well-drained soil, try *Zoysia tenuifolia*. This grass will not give you a conventional carpet of lawn, but will spread to look like an undulating mat and is said to never need mowing. It is sometimes available from nurseries in small pots. A new release of zoysia is *Zoysia japonica* 'Empire', which is said to suit a wider climatic range, be salt-tolerant and drought-tolerant.

- Begin the same way as for sheet mulching by closely mowing or brush-cutting the area. Weed control is vital, so you might prefer to dig over the area to remove the weeds, or to solarise it.
- Spread any necessary soil additives.
- Cover the whole area with a thick layer of well-overlapped newspaper or cardboard, or a commercial weed mat. I used newspaper and now wish I had spent the money on a manufactured product as I had a very persistent creeping grass that kept growing through the paper. Undaunted, I pulled it out and re-covered the spots. My persistence won and the grass, being deprived of sunlight for photosynthesis, ultimately died beneath the newspaper.
- Cover the base weed mat with a cosmetic layer such as pea straw.
- Leave the area alone for as long as your patience allows, so the weeds will die and decompose underneath. Be vigilant about removing any weeds that grow through the sheet mulch and re-covering the spot they came through.

- When you are ready to plant, cut or tear a hole in the base weed mat and plant into it. I used pennyroyal, creeping thyme and lawn chamomile. Other possibilities are lippia, creeping boobialla, kidney weed, Corsican mint or native violet (for shady, low-traffic areas only).
- Keep plants moist until established. The mulch layer ensures that little water is needed. Depending on the time of year, a drink once or twice a week should suffice.
- The herbs spread over the mulch in time. The mulch decomposed to add organic matter to the soil and reduce the need to water. Mowing, with the blades on a high setting, is needed infrequently, and we have a fragrant floral carpet in our garden room. After mowing, especially in summer, a pennyroyal herb lawn can look dead for a time. Don't despair. The roots are still in the soil and it won't be long before new growth appears.
- The same technique can be used to establish a lawn of native grasses, or a wildflower meadow.

This is a weed control method for patient gardeners. For best results, solarise garden beds or areas designated for lawns in the hottest part of the year.

- Closely cut the area.
- Some garden writers recommend applying fertiliser to stimulate quick, sappy weed growth. This is optional.
- Water well.
- Cover the area with thick plastic and seal the edges with weights, even soil will do. Clear plastic is usually recommended in garden books; however, a trial of organic gardening methods in Tasmania found black plastic to be more effective.
- Leave for at least six weeks so the high temperatures under the plastic kill all weeds and seeds.
- Remove the plastic and proceed with soil preparation as discussed in the previous chapter.
- Variations of temperature mean that this is not an exact science. If the temperature hasn't risen enough, some weeds and seeds will have survived the treatment. If the temperature has risen too high, beneficial soil biota will also have been destroyed. Sheet mulch will stifle any remaining weeds and compost will reintroduce beneficial biota.

Individual plants

There might be some spots you want plants that are not part of a garden bed, perhaps a specimen tree, a copse or some shrubs on the nature strip. My favourite technique for ensuring these plants thrive is what I call the mound, moat and mulch method.

The plant is in a slight mound surrounded by a moat.

You can see how the moat prevents water runoff.

Applying mulch around the plant.

The mulch is spread over the mound and the moat.

The method has been used for several plants on the nature strip.

- Dig a large hole where the plant is to go. If your soil is compacted, clayey or just generally in poor condition, forget the oft-repeated advice to dig the hole twice the volume of the pot the plant is in. That might serve well enough when the soil is loamy and fertile, but better preparation is needed if your soil is not ideal. My advice for anyone with less than ideal soil is to dig the hole at least four times the volume of the pot.

- Dig in any combination of compost, rotted manure, blood and bone, Dynamic Lifter or other soil conditioner, along with any amendments your soil needs in the way of gypsum or lime. I like to use a double handful each of compost and rotted manure per hole – the larger the hole and the poorer the soil, the more you need – then I can sprinkle blood and bone or other organic fertiliser on the surface at a later time. In areas of very hard clay I've added up to a bucket of bought topsoil to the mix. Remember, most native and hardy Mediterranean plants do not need a lot of fertiliser, especially fertiliser containing phosphate, so tailor what you work into the soil to suit plant needs and the fertility of your existing soil.

- Mix all additives well with the soil from the hole.

- Refill the hole to a level that will allow the new plant to go in so it is at the same level in the soil as it was in the pot.

- Put the plant in the hole, loosening its roots a little first if needed.

- Backfill the hole and form a moat around the plant with the excess soil.

- Fill the moat with water.

- Spread a thick layer of mulch over the dug area and moat, but not right up against the stem of the plant, and dampen it well.

- Subsequent waterings will soak the mulch and fill the moat, ensuring that your plant is efficiently watered and the soil stays damp for an extended period.

- The thorough soil preparation will enable the roots of the new plant to easily move through and reach all the goodies you added to the soil.

By now you probably have drainage installed where necessary, hard surfaces laid, structures erected or at least allowed for and soil well prepared, just waiting for the most exciting part of any garden – planting time.

PLANTING TIME

For me, plants are the reason for gardening; everything else plays a supporting role to help display the stars to their best advantage. Apart from their intrinsic attractions, plants are essential to the support of vast networks of biodiversity. Careful choice of plants, as detailed in earlier chapters, can transform your garden into a beautiful oasis of biodiversity. The biodiversity benefits will be multiplied if your garden can link up with other gardens and public reserves to form larger, more sustainable biolinks.

With plants ready in their pots and garden beds prepared, it's time for some fun.

When to plant

To ensure less stress on new plants and increase their chances of thriving, there are some times of year, times of day and weather conditions that are the best planting times.

Time of year

If you can, plant in the season that has the mildest weather conditions. Here, that means either just after the autumn break when the soil is still warm and the rains have started, or once the soil has warmed up a little in spring. Winter planting in areas where this season is not too severe can be successful, but plants will be under stress if they are subject to frost or waterlogging, and will not thrive as those planted in a milder season will. Don't plant in the heat of summer unless you can keep plants well watered and provide shade. If you do plant in summer, try to do so at the start of a cool change so new plants have at least a few days of milder temperatures before having to cope with extreme heat. During drought years, wait patiently for the weather to break and use the time for planning and research.

Time of day

Evening is a good time to plant, giving plants the hours of darkness and coolness to recover from the stress associated with removal from the pot, root disturbance and relocation.

Weather conditions

Try to wait for a cloudy day. Rain is even better if you don't mind getting wet in a good cause. In cloudy or rainy weather you can plant at any time of day. Do avoid planting in the heat of the sun. Even in spring and autumn the sun is hot in the middle of the day and plants will be stressed, so if the designated planting day is sunny, wait until evening.

Plant placement

You might have a very clear idea in your head, or on paper, of exactly where in each bed each plant is to go. However, in three dimensions in the real garden, things can look different. It is always wise to lay out the pots where you intend them to go. Work on one area or bed at a time so you are creating a complete picture.

Colour schemes

Have you decided on contrasting or toning colours, a single colour or a rainbow mixture for this area? Ideally, all your potted plants would be in flower when you planted them so you could judge whether or not the effect was what you were after. In reality, this is unlikely to be the case, so you will need to rely on some vivid mental imagery and the colours shown on the label. Most illustrated plant labels give a reasonably accurate idea of the

The silver birch in this bed will grow tall to shade and drop free mulch on the understorey.

flower colour you can expect, but some depict exaggerated colours. This is where time spent looking at magazines and a variety of other gardens pays off.

Remember that plants will flower for a limited time of the year. The structure of the plant and the leaf colours will be important for the remaining time. Do you want all plants in this area to flower at the same time, or have you planned for staggered flowering so there is something flowering in the bed at most times of the year?

Remember that all-white or all-pastel gardens have limited biodiversity value, so plan a more vibrant colour palette for at least one area of the garden.

Size

Plant labels should indicate the mature width and height of each plant. You must allow for this in spacing out your pots. If they seem too far apart and the garden looks bare, you can fill in with annuals or biennials in the short

Tall trees with shrubs in the understorey at Canberra Botanic Gardens.

The small shiny leaves of the 'Orange Twists' contrast with the strappy irises.

Phormium 'Yellow Wave' contrasts in colour and texture with lavender.

term. Why not consider growing some edible herbs among the shrubs for as long as there is space? Many herb flowers are great attractants for butterflies and other beneficial insects.

Layers

A layered planting scheme has many advantages. The taller plants shade and drop leaves to mulch those lower down; medium plants fill in between the taller ones and also shade and mulch the lowest storey. The ground-covering lowest storey acts as living mulch for everything above it and provides interest and colour at ground level. This scheme provides the maximum habitat niches for a range of fauna and is very water efficient.

Textures

Textural variety of leaves, flowers and bark provides visual interest at different times of the year. Combine strappy leaves with round; silvery, furry leaves with dark glossy ones; large deltoid or oval leaves with graceful grass blades.

Compatibility of care

To minimise long-term maintenance and maximise water efficiency, group in the same bed or area plants having similar water and fertiliser needs. Quite simply, don't plant ferns and fuchsias together with hardy natives.

Island beds

Island beds can be walked around and admired from any angle. A good planting layout for these beds is to have permanent

Increase the 'wow' factor with variety of leaf colour, shape and texture.

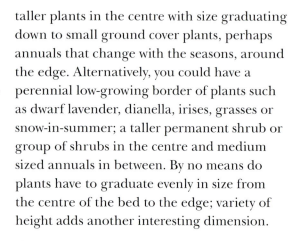

An example of gradation of height beginning to develop in an island bed.

A pretty example of a mixed border that separates the garden from the paddock.

taller plants in the centre with size graduating down to small ground cover plants, perhaps annuals that change with the seasons, around the edge. Alternatively, you could have a perennial low-growing border of plants such as dwarf lavender, dianella, irises, grasses or snow-in-summer; a taller permanent shrub or group of shrubs in the centre and medium sized annuals in between. By no means do plants have to graduate evenly in size from the centre of the bed to the edge; variety of height adds another interesting dimension.

If your preference is for a bed planted with permanent native shrubs, for example, you might like to incorporate seasonal changes by interspersing a variety of annual flowers or herbs.

Borders

A border might be a bed planted along an existing fence or it might act as a fence substitute. For a border along a fence, plant taller shrubs or small trees at the back, along the fence line, graduating to smallest at the

front. If a border is used as a fence or partial fence, follow a planting design similar to that of an island bed. Have taller plants running along the length in the centre of the bed, graduating to smaller on each side, so the view from the road and from inside the garden is equally attractive. Of course, it need not look the same from each side.

A narrow border, between two houses for example, might consist of a row of permanent screening plants with herbaceous perennials in between and low-growing ground covers, either annuals or perennials, spreading at the base.

Copses

I always feel sorry for the single specimen tree in the middle of the lawn, all by itself with no plants around it to ameliorate winds and harsh sun or to help attract birds and insects. Try a small copse instead, a group of three or five plants, either the same species or a variety of species with similar care requirements. Place the copse so it will shade

Pansies (annuals) and alpine strawberries (perennials) in a border beneath the palm trees (perennials).

Perennial shrubs and strappy plants.

a veranda or recreation area if you can. A birdbath and a seat in the shade of the trees make a welcome spot for humans and feathered species to frequent.

Annuals

Annuals usually need to be replaced each season and their water needs are often high, making them high-maintenance plants, but they do add colour and variety to the garden and can fill in bare spots while permanent plants grow. Intersperse them with permanent shrubs and trees or use them as border plants.

Perennials

The term perennial can mean a plant that lives for more than three years, including most trees and shrubs. It is often used, however, to refer to herbaceous perennials that die back each winter to a rootstock, rhizome or tuber and grow again each spring, usually flowering in late spring or summer. Bulbs are herbaceous perennials and,

depending on the species, can flower from late winter through to summer. Unlike annuals, most herbaceous perennials don't need to be replanted each year. They often have spectacular flowers and are worth interspersing among your permanent shrubs and trees for seasonal colour surprises and cut flowers for the house. Do remember where they are. Maybe mark them in some way, so you don't try to plant into the spots when they are dormant.

Focal points

Try to lay out each bed so it has a focal point or feature that attracts the eye. View the bed from the angle it will most frequently be seen – maybe from a house window, from the street, your entrance gate or outdoor living area. Rearrange your pots until you are satisfied that there is a plant or a grouping placed so it will draw the eye. It might be the tallest plant in the bed, or one with an interesting shape, a stunning floral display or unusual textured leaves. The focal plant or

Arctosis around a stump makes a stunning focal point when in flower.

Irises in the island bed are a focal point when flowering.

group should attract attention, but still harmonise with other plants in the bed.

A focal point might not be a plant at all. It might be a statue, garden art or water feature, either used alone or in conjunction with a plant or group of plants.

The focal point might change with each season. Maybe the stunning floral display of a callistemon in the centre of a bed will be the feature for a short while in late spring, whereas in winter and early spring the more subtle tones of thryptomenes contrasting with splashes of colour from daffodils or jonquils creates a different focus.

On the wild side

In all but the most formal garden designs there is opportunity to let plants go a little

This grouping of roses, box hedge and weather vane is especially eye-catching when the roses are in bloom.

In the same bed slightly later in the year, poppy flowers grab attention.

wild. Allow room for annuals and herbaceous perennials to set seed and germinate where they will. Or encourage the process by cutting off seed heads and shaking seeds in other areas. There will be plenty of space between the permanent plants in the first few years and by the time the permanents are spreading and taking up more room, the self-seeders will have naturalised enough to seed down in any spare niches. Some plants that successfully go wild for me include: calendula, columbine, borage, nasturtium, poppy, hollyhock and erigeron. Even in the vegie garden I encourage parsley, rocket and celery to self-sow by using plants with mature seed heads to mulch other beds.

Proximity to house

Beds adjacent to the house, especially in front of windows, present some design challenges. Careful selection and placing of species is needed to ensure that plants in front of windows are placed to look their best from inside the house, but do not grow large

enough to obstruct the view or block light. On the other hand, if you want to shade windows from westerly sun in the summer, you might choose medium-sized deciduous plants so you can have summer shade and winter sun. To get an idea of where the shade will fall, get someone to stand holding a long stick at arm's length above their head where you want the tree to be and watch where the shadow falls.

Please, do not plant tall trees too close to the house.

For energy efficiency, mainly your own, place higher maintenance species and those needing more water closer to the house, and tougher low-maintenance species further away. Make sure any plants that need shade or dampness are positioned in the appropriate spots. To ensure they thrive and to reduce water loss through transpiration, the more tender plants also need to be in areas with some wind shelter, provided by the house or shed or by hardier species positioned as windbreaks. Remember that young windbreak trees will not perform that role effectively for a number of years so you might need a temporary screen of some sort.

It might seem as if there is a lot to think of when positioning plants, but most of it is just common sense. By the time you have planted a bed or three it will be second nature.

In the ground they go

At this stage you should have beds and planting holes prepared, as described in Chapter 5, and pots placed where you have finally decided to plant them. Here's how it's done.

- Water the pots and the holes or beds plants are to go into.
- Press firmly all around the outside of the pot to loosen the soil. If a plant has been in its pot for a long time there might be roots growing out of the drainage holes and the pot might be more difficult to remove. In the case of protruding roots, cut them off with secateurs. If a plant is difficult to remove, try tapping all around the pot with the flat back of a trowel. If it still won't come out, put the pot on its side on the ground and roll it firmly with your foot. The final extreme solution is to cut the pot with a sharp blade.
- Holding the pot in one hand, tip it upside down so the emerging plant is supported in the palm of your other hand. If a large plant is difficult to remove in this way, having loosened the soil as described above, lay the pot on its side on the ground and have someone hold the pot by the base while you gently but firmly pull the plant out. If you think the plant is going to be damaged, it is best to cut the pot.
- Sometimes a plant does not have a very well-developed root system and is potted in a loose mixture that falls away easily. Support the roots of such a plant with cupped hands and gently place it in the hole.
- Sometimes the roots are quite tightly bound in the potting mix, perhaps even running around and around inside the pot. In this case it is necessary to loosen the roots somewhat before planting so they will be able to move out into the soil of the garden bed. Try loosening the roots gently with your fingers first. If they are too tightly bound for this to work, it might be

necessary to cut through the root mass vertically at several places around the circumference with a sharp blade. Cut off any damaged roots with secateurs before planting. Alternatively, soaking the root mass in a bucket of water for several hours might loosen it.
- Position the plant in the prepared hole so the soil comes to the same level on the stem as it did in the pot.
- Backfill the hole with the soil you dug out of it.
- Press firmly all around to tamp soil down and support the roots.
- If planting individual trees or shrubs, form a watering moat as described in Chapter 5.
- If any roots have been damaged or cut during planting, it is best to prune a similar amount from the top growth of the plant because the reduced root mass will not be able to provide nutrients for the whole plant.
- Plants will grow a thicker, stronger stem if they are not staked, but if a plant is tall or top heavy it will need some temporary support. This is best provided by tying the plant to the stake with a length of old pantyhose or similar so that its stem is enclosed by a figure of eight. For larger plants a stake on each side might be necessary. The stem is able to move within the support, and is thus stimulated to grow a thicker trunk, but cannot be blown over.
- Water your new garden.
- Stand back and admire.

Mulching

No garden bed is complete until it's mulched.

Benefits of mulch

Mulch is amazing stuff and benefits your garden in so many ways that you can't afford to omit this finishing touch.

- The most important value of mulch is in water conservation. Mulch slows evaporation by as much as 80%. Moisture remains in the soil for longer and you have to water the garden less frequently. You will reduce your water bill and be helping conserve water. Organic mulches decompose and are incorporated into the soil, increasing its absorbency so it holds more moisture. Myriad creatures live in and under the mulch: earthworms, spiders, ants, millipedes, nematodes, mites, termites, springtails, beetles and numerous other visible and microscopic organisms. Many of these organisms tunnel into the soil, leaving air channels that allow better water penetration to deep levels of the soil. In addition, the soil biota eat each other, defecate and die in the soil, a process that increases nutrients and organic matter.

This copse has been mulched with chopped timber and leaves from a tree lopper.

- In well-mulched soil there is a high level of organic matter that encourages a larger number of beneficial micro-organisms and discourages harmful ones. This also increases disease resistance.
- As organic mulch decomposes it releases nutrients that enrich the soil and are continuously available to plant roots. This reduces the amount of fertiliser you will need to buy. It also reduces the amount of excess nutrients entering the ecosystem that can be leached into water tables, lakes and rivers.
- The slow decay and incorporation of mulch into the soil improves the structure of both sandy and clay soils. Sandy soils hold more water and nutrients and clay soils have improved drainage and better availability of nutrients.
- The mulch layer insulates the soil and maintains a more even temperature in the root zone. This means less stress for plants and healthier growth both above and below ground.
- A layer of mulch over the soil helps prevent compaction from foot traffic and heavy rain and erosion from rain and wind.
- The more variety of mulches you use over time, the more nutrients are released on their decomposition and the more you help the soil develop and maintain a garden-friendly stable pH that enables minerals and trace elements to be available to plant roots.
- Mulch reduces weed growth and usually allows weeds that do grow through to be easily removed.
- Most organic mulches are by-products and wastes from farming, forestry and food production and it is satisfying to know we

can put them to such good use in our gardens instead of having them go to environmentally damaging landfill.

- Much of the waste from our own kitchens and gardens can be reused as mulch, perhaps via the compost bin – another saving for our pockets and for the environment.

What to use

The variety of mulches available is truly amazing, from commercial packaged products, to industrial wastes you can usually obtain free of cost, to what you can grow or recycle at home. When starting a new garden you will not be producing prunings, compost, spent annuals, grass clippings or fallen leaves and leaf mould to use as mulch for some time, so the first mulches you use will more than likely be bought ones. There are numerous packaged materials available at nurseries. Some of my favourites are organic sugar cane mulch, pea straw, lucerne hay, decomposed manures and compost. I do not like, nor do I recommend, pine bark, unless it is used as a path. It contains resins that can inhibit plant growth, takes forever to decompose and adds little in the way of nutrients – indeed, it can rob nutrients from the soil as it decomposes. Wood chips I rate as only slightly better than pine bark and only use them myself around hardy native plants.

Nurseries and landscape suppliers will stock a range of bulk mulches that could be cheaper than buying packaged materials in smaller quantities, but check the delivery cost.

If you are lucky enough to come across a tree lopping contractor with a mixed load of chopped timber and leaves to dispose of (not

Newspaper sheet mulch beneath pea straw.

pine or cypress trees), this makes a very good mulch for native gardens. We were able to obtain a truckload for the cost of a slab of beer.

Check out your council's green recycling system. Often mulch or compost is available for little or no cost if you collect it yourself. Also check any food production industries in your area. There might be wastes such as nut shells, pea trash, bagasse (sugar cane trash), grape marc, apple pomace, rice hulls or brewery waste available to anyone willing to take it away.

For sheet mulching you can use recycled newspaper or cardboard (the very thick boxes from televisions and refrigerators are terrific) or buy one of the commercial weed mats.

What to use where

Two good general-purpose mulches are pea straw and chopped sugar cane mulch. You can use coarser materials, such as shredded trees, wood chips and leaves, and less compost and manure on native plants and other hardy shrubs and trees. Use more finely chopped

Copse being sheet-mulched with cardboard and shredded trees.

Pebble mulch over commercial weed mat.

materials and more compost and manure on annual flower beds. Lucerne mulch is good to use in the vegie garden and for roses.

Pebbles, scoria, rolled coloured glass and similar hard materials are very suitable for mulching succulent gardens and the surrounds of water features. Do ensure you use a weed mat or thick cardboard underneath them – it is a very fiddly, fingernail-breaking job to remove weeds coming through these hard mulches. For small areas such as around a water feature you can buy large bags of pebbles from most nurseries and hardware suppliers. If you have a larger area to cover, try your local landscape supplier or a nearby quarry.

What to do

Simply scatter your chosen mulch all over the damp bed, or around individual plants, to a depth of 15 cm for coarse mulch, 10 cm for medium coarse or 8 cm for finer materials. Take care not to let mulch build up directly around plant stems or trunks as a fungal infection known as collar rot might result. Some plants are very susceptible to collar rot, while others, including most herbaceous perennials, will grow right up through thick mulch without any problems. I like to scatter a thin layer of compost or decomposed manure over the bed – as fertiliser and mulch – before adding the top layer of mulch; this helps the mulch decompose and provides nutrients that are probably not in the mulch.

Once mulch is applied, water it well. Damp mulch applied to damp soil gives best results; it keeps the soil, and hence plant roots, moist and creates a favourable environment for beneficial soil and surface biota.

To maintain soil health and maximise the benefits of mulch, vary the material used each time.

You're making progress

Well done! You now have at least one plot planted, mulched and watered. Pat yourself on the back, enjoy a cool drink while you admire your creation, and then have some fun adding decorative touches – statues, rocks, coloured pots, an obelisk, or any form

ACTION PLAN

- Choose the time to plant.
- Lay out plants in pots and adjust the planting design.
- Remove plants from pots and place in prepared holes.
- Backfill and tamp soil down.
- Prune and stake if necessary.
- Water.
- Apply mulch.
- Finish off with decorative touches.

of garden art that complements your theme. Now proceed to create other beds using the same steps. You'll find the second and subsequent beds much easier now you know how to go about it. The next thing you need to think about, if you haven't done so already, is how you are going to water each bed.

WATERWISE OPTIONS

Anything from a third to a half of the water consumed by an average household is used on the garden, so the way we design and water our gardens can have a huge impact on our water use. Previous chapters have detailed a variety of water-efficient strategies, and you've probably already incorporated many of them by this stage.

Checklist for water efficient design

Water efficiency in the garden involves much more than the type of watering system(s) you install. This brief roundup of points mentioned in earlier chapters will allow you to make sure your total strategy for water efficiency is in place.

- Choose plants that will tolerate dry conditions (Chapter 4).
- Add organic matter to the soil to increase its absorbency (Chapter 5).
- Aerate compacted soil (Chapter 5).
- Position plants appropriately to their needs (Chapter 6).
- Plant in layers (Chapter 6).
- Group plants with similar watering needs (Chapter 5).
- Use mulch (Chapter 6).

When to water

Watering times in most areas are usually determined by the permanent restrictions that apply. Avoid watering, even by hand, in the heat of the day, unless the area being watered is in shade. Avoid using sprinklers, as far as possible, during extremely windy weather. Do not water every day as a matter of course. Test the dampness of the soil by scratching under the mulch in a friable spot with your finger, or by digging with a trowel to the depth of its blade. If the soil feels damp, don't water. You'll soon work out a system and know how often each bed needs watering. In time, you will be able to tell by the amount of leaf droop at the end of a hot day which beds or individual plants need water.

For the first summer, newly planted beds will probably need to be watered at least twice a week. Seedlings will need watering every day, maybe even twice a day in hot weather, until they become established and hardened, then gradually extend the period between watering them. The vegie garden will possibly need watering every two or three days once seedlings are established, depending on heat, wind, amount and type of mulch, and microclimate.

Even beds of permanent waterwise trees, shrubs and understorey plants will need regular watering during the first two to three years. Over time, the plants will put down deeper roots and become acclimatised, the mulch you have applied will continuously decompose and be incorporated into the soil to increase its water-holding capacity, ground cover plants will spread to keep the soil cool and moist, and the plants will grow to provide shade and wind filtering for each other. All this progress means less frequent watering is needed. You should be able to extend the period between waterings to about two weeks by the end of the second summer for areas of native and similarly hardy species.

My aim, with areas of native plants, was to limit them to one or two deep soakings per summer, to supplement rainfall, by the end of the fourth summer. A severe drought during the fourth summer ensured that some areas received no extra watering. In my previous garden on a country property where we depended on rain and dam water, I watered some areas by bucket for the first summer only. After that, anything that didn't survive was replaced by something that did. I lost a lot of plants (and money) at first, so it took a while for some areas to become established, but they were eventually very drought-tolerant. You might prefer to adopt a similar hard-nosed attitude. If so, be prepared for some trial and error to determine the most drought-tolerant species for your area.

Watering systems

There are numerous choices when it comes to watering systems, everything from simple

A range of pipe and fittings to install a drip irrigation system.

hand-watering to sophisticated automatic microspray or dripper systems. As with most choices, you need to strike a balance between maximum efficiency, affordability, convenience and appropriateness for the type of bed.

Hand-watering

Watering by hand with a trigger nozzle on the end of a hose is a cheap option and can be a good temporary measure if finances do not allow the installation of an irrigation system at this time.

Advantages

Hand-watering is cheap, requires no special installation and you can direct the water exactly where it is needed so there is little or no waste. It gives you the opportunity to be constantly aware of what is happening in the garden, so you know when jobs such as weeding, mulch replacement, pruning or pest control (organic methods only please) are needed. You know, too, when the first flower buds are opening, when honeyeaters or butterflies are visiting and which plants they prefer. You can also be doing a little gentle weeding around plants at the same time.

Disadvantages

Watering this way is time-consuming and, unless your soil is friable and well aerated, water will run off instead of soaking in. See the mound, moat and mulch method described in Chapter 5 for a means of ensuring maximum efficiency when hand-watering. You need to be patient enough to make sure each plant is watered deeply, otherwise shallow watering will result in shallow-rooted plants that are subject to wind damage and heat stress and will need more frequent watering.

Sprinklers

The traditional overhead sprinkler that throws a fine spray of water into the air might be cheap to buy, but is costly in terms of wasted water. The higher the water is thrown and the finer the droplets, the more water is wasted. All too often such sprinklers are used on windy evenings when much of the water is blown away, or they are left in place for too long, resulting in water running off into gutters. There is also the far too often seen scenario of sprinkler overspray watering paths, driveways and even the road.

For lawn watering a system of pop-up sprinklers can be installed. These are permanently in place and cannot be seen when not in use. The feeder pipe, black plastic polypipe, needs to be buried in the soil before the lawn is planted. The sprinklers are inserted at the required intervals. When the tap the system connects to is turned off, the sprinklers retract into the pipe; turn the tap on and the water pressure causes the sprinkler to rise. Pop-up sprinklers are available in a range of spray patterns and there is even now an insert that is claimed to reduce their water use by half. Talk to your irrigation supplier about installation and what will best suit your needs.

Permanent fixed sprinklers for garden beds are installed in a similar way to pop-ups.

The first few times you use any sprinkler on a particular spot, check frequently to see how far down the soil is damp and when runoff occurs. This will give you an idea of how long to leave a sprinkler in one spot before moving it. I find about 15–20 minutes for most beds is adequate. If runoff occurs before the water has soaked in adequately, your soil is compacted. Aerate it with a fork and keep adding organic matter.

Advantages

Most sprinklers are cheap to buy (though a pop-up or other fixed system will cost more) and, if used carefully, are a useful stopgap or backup. Apart from pop-up or other permanent fixed systems, they require no special installation. Any type of sprinkler creates a humid microclimate around the watered area that is of benefit to many plants, especially in very hot, dry weather. Sprinklers also wash the dust off leaves, encouraging more efficient photosynthesis, and help dislodge or discourage some pests. Look for sprinklers that will water a variety of different shapes, use them on wind-free evenings or

Polypipe being laid for a microspray system.

early mornings, and don't forget they are on. There is now a water-efficiency rating for sprinklers, so look for this label when buying.

Disadvantages

The biggest disadvantage of sprinklers is their wastefulness, both of your money when it comes time to pay the water bill, and of a limited resource. Portable sprinklers need to be moved regularly from one spot to the next, so constant supervision and awareness are needed.

Microspray system

A microspray system consists of a length of black plastic feeder pipe, fixed more or less permanently around the bed to be watered, into which are inserted tiny sprinklers – hence, microspray. The sprinkler fittings come in an assortment of spray patterns to suit the shape of the bed and can be mounted on black plastic risers of varying heights. The feeder pipe connects to a tap and usually feeds through a filter, attached close to the tap, to remove particles that might otherwise block the microsprays. An automatic timer device is an optional extra.

The feeder pipe can be laid so it wends its way through a bed, or it can be attached to the edging. It can be placed prior to the mulch

being applied, but it is simple enough to scrape existing mulch out of the way, lay the pipe, then spread mulch over to cover it.

Advantages

Microsprays direct the water pretty much where you want it, though they are subject to some wind-borne overspray. Water use is quite efficient and once you have determined the optimum watering time for each bed, they

Tap to control water flow to each bed or section.

Microspray system in operation.

require no supervision, apart from remembering to turn them off.

Disadvantages

There is some water wastage during windy weather. A system can be costly to set up, perhaps as much as several hundred dollars, depending on the size of your garden. Installation is required, but this is within the capability of the average handyman. Plants growing directly in front of the microspray can block the water from plants further back. If this happens, you can install taller risers or adapt the planting design to ensure there is nothing growing in front of the microspray.

Dripper systems

A dripper system works in much the same way as a microspray system, except that drippers replace the sprayers. The dripper devices are available in a variety of styles and most have an adjustable flow rate. They can be inserted directly into the polypipe or on lengths of flexible spaghetti tubing attached to the polypipe. Some dripper tubes have holes at set intervals along the length. Talk to your irrigation supplier to determine the style most suited to your needs.

Both pipe and drippers are best laid under the mulch.

Advantages

This is probably the most water-efficient system. There is no loss to wind or evaporation and water is directed at the root zone. No supervision is required during use, but they do need to be checked periodically to ensure there are no blockages.

Disadvantages

A dripper system is similar in cost to a microspray system. Drippers can easily become blocked with particles or sediment, and are fiddly to clear. The mulch is not kept damp, so does not have the same benefits of decomposition, soil enrichment and encouragement of beneficial soil biota as damp mulch does. Many plants benefit from the humid environment provided by sprinklers and microsprays and this is lacking with drippers. There is no water spray to clean the leaves of dust or to dislodge pests. If plant roots do not reach the drip zone, they get no water, so supplementary watering is needed for seedlings and small plants until their roots spread.

Seeper hose

In theory this product sounds ideal. It is a porous hose made from recycled tyres, and is laid meandering around or along the centre of garden beds, covered with mulch and attached to a hose by a snap fitting. When the

Tap controlling water flow in a length of seeper hose which goes under the mulch.

tap is turned on, water seeps out of the seeper hose right into the root zone where it is needed. Perfect, a recycled, water-efficient product! Maybe.

Advantages

The hose is simple to install and needs no drippers, microsprays or filters. It is completely covered, so will not spoil the aesthetics of your garden. It is probably not as water-efficient as a dripper system, but almost, with no loss to wind or evaporation. It is difficult to compare prices of seeper hose and a dripper or microspray system because there are so many different options and combinations available. However, as seeper hose can usually be installed without the need for extra fittings such as risers, spray nozzles and feeder tube, it is probably cheaper than an equivalent dripper or microspray system and very similar in cost to the type of drip tube, referred to below under *Gadgets, gizmos and good ideas*, that merely has holes along the length.

Disadvantages

With the hose I bought, watering is very uneven. A lot of water seeps out close to the tap and progressively less along the length of the hose. A length of any more than about eight metres means that plants at the end of the run receive little if any water. I had to adapt the installation so as to have several shorter lengths of seeper hose attached at right angles to a length of feeder polypipe with a tap at each junction so each length of seeper hose can be turned on or off separately. Even a slight uphill slope or a hump in the ground decreases the flow and thus the viable length of hose. Plants need to be close to the hose to access the water, so newly planted areas might need supplementary watering until their roots spread. The same comments about humidity, mulch effectiveness, dust and pest removal that applied to dripper systems also apply to seeper hose. In time, it is possible, as one plumber cautioned me, that plant roots will grow into and block the hose.

Soaker hose

Soaker hose is a semi-transparent green hose with tiny holes punched at intervals along its length. It can be used with the holes facing up as a long sprinkler, or with the holes facing down as a seeper hose. It is most conveniently left *in situ* and attached to a normal hose by a snap fitting.

Advantages

It is cheap to buy and good for watering long narrow beds. It is quite flexible, so can also meander around a bed of almost any shape, but will not cope with sharp angles.

INSTALLING A DRIP OR MICROSPRAY SYSTEM

Consult the staff at specialist irrigation supply outlets or look for the manufacturer's brochures at general hardware stores before deciding on and designing your system. There are starter kits available at hardware stores and nurseries that are quite economical and might suit your needs.

The simplest method to use is the soft poly method – thin-walled black flexible polypipe that is easy to bend and cut; the fittings press on easily and are then clamped into place.

- Start with 19 mm diameter pipe as a trunk system from the tap. This will be the main from which you can attach smaller branches and taps and is connected to your normal garden tap with a plastic screw fitting (nut and tail).
- For branches use 13 mm pipe with individual taps at each branch to control flow. These connect to the main pipe with tee-fittings.
- The branch pipes follow the general line of the garden bed. If you can design your system to bring the branch around in a full circle connected back into itself, the circuit will allow a full flow without loss of pressure.
- Connect dripper or microspray devices directly into the branch pipe, or attach 3 mm feeder pipes leading directly to each plant, with an adjustable dripper on the end of each feeder pipe.
- An added advantage of the soft poly system is that you can change the layout and location of the water outlets simply by using so-called 'goof' plugs which plug up the hole you have made in the pipe.
- This basic system can be developed further by the installation of timer devices programmed to open up the taps at certain times, thus making the system fully automatic. If you go down this path, do remember to install an override to prevent the taps from opening when the soil is already damp.

Disadvantages

As with any sprinkler, it is subject to water loss by wind and evaporation. Some of the perforations allow water to escape at odd angles, completely missing the target zone.

Gadgets, gizmos and good ideas

Water-saving products are set to be one of the growth industries of the 21st century and there are some nifty devices around. There are also some innovative ideas that won't cost you anything but a bit of time and some rubbish crying out to be reused. Look around large hardware stores and you'll find plenty of water-saving toys to play with; below are just some of them.

Purpose-made tools and fittings for irrigation systems

There are numerous clever tools and fittings to make installation of a drip or microspray irrigation system easier, such as the device that punches holes in the feeder pipe for the microspray or riser to fit into. When buying the components for your system ask about relevant tools. They are usually quite cheap to buy and save much time and sweat.

Trigger nozzles

A trigger nozzle attached to your hose with a snap fitting lets you stop the water flow when moving between areas. These nozzles come with adjustable flow rates or a variety of spray patterns.

Multiple taps

Available in a variety of styles, dual, triple or quadruple tap adapters connect to a single tap to allow you to use a hose and a watering system, or multiple systems, from the one outlet. The simple option is a Y-shaped device that screws onto a threaded garden tap and has a switch on each side to direct the flow. There is a very sophisticated battery-operated device that acts as a timer unit as well.

Water-saving hose

Looking like normal hose, but very much more costly, water-saving hose has a valve fitted inside it to maintain a constant flow at varying water pressures.

Drip tube

There are at least a couple of different options when it comes to drip tubes. One product is black plastic tubing with holes spaced at intervals of 10 cm, which attaches to a normal hose by a snap fitting. The discharge rate of a litre per hour, under low water pressure, is not adjustable with this tubing.

Moisture sensors

Used in conjunction with an automatic timer, one type of moisture sensor is buried in the ground, detects soil moisture and stops the preset watering cycle until the soil dries out.

A more sophisticated version, which might need professional installation, monitors evaporation and rainfall and determines when plants need watering. Known as the Aquamiser™, this one might not be available in retail outlets; to find out about it visit: www.betterwateringsolutions.com.au.

Watering spike and tubes

A watering spike is simply a long rigid pipe with a spike at one end that is inserted into the soil near a shrub or tree. A hose attaches to the top; water travels down the tube, emerging from holes in the in-ground section, to be delivered to the root zone. There is a similar tube-shaped device as well.

Tanks

Tanks are now available in a range of sizes, shapes and colours to suit most garden applications. There is a very innovative design available in the shape of a fence or wall 330 mm deep, 1800 mm high and 2400 mm long. It holds 1200 litres and several can be joined together to increase storage capacity.

Pressure-reduction valves

If your water pressure is very high, chances are you are using too much water on the garden, and elsewhere. You might also have a problem with joins in irrigation systems blowing apart under pressure. A pressure-reduction valve fitted at the water meter will reduce the flow and save water and your money.

Look for the lilac

Lilac is the internationally recognised colour for grey water pipes, hoses, sprinklers and

Lilac is the internationally recognised colour used for grey water plumbing devices.

other fittings. It is a very practical and appealing way to differentiate equipment used for recycled water only (see below).

Homemade and cost-free

To water individual plants you can use a large soft drink or juice bottle. Cut a hole in the bottom to allow hose access and insert the open top into the soil beside a plant. Fill the bottle with water through the hole in the bottom (now at the top). A plug of coconut fibre can be placed in the neck of the bottle before putting it in the soil to prevent the opening from clogging up. Check occasionally and replace coconut fibre when needed.

A similar idea uses any wide-diameter rigid pipe that might have been left over from the plumbing of your home. Dig a hole, or several, beside the plant at a distance of about 30 cm and insert the pipe, angled slightly towards the plant, to a depth of about 25–35 cm. Cut the pipe so it is just visible above the soil. With mulch around it you won't even see it – there's no need for it to

look an eyesore. Fill pipe(s) with water when needed. The water seeps into the soil directly into the root zone. In very wet weather when the soil is already saturated you can cover the top of the pipe.

Grey water

Grey water is used water from the bathroom and laundry. Kitchen water with its larger complement of solids, cleaners and fats is considered dark grey and is unsuitable for reuse in the garden. There are numerous methods of using grey water in the garden, from the simple but back-breaking task of bucketing out the bath water onto favourite plants, to systems that filter and treat the water then automatically irrigate the designated areas.

Diversion system

The simplest system uses a diverter switch mechanism that intercepts the water along an existing discharge pipe and directs it to the garden by gravity. Look for a diverter switch that permits the grey water to resume flowing into the sewer in wet weather when it is not needed on the garden or if blockages occur.

Diversion and filtration system

The addition of a filter mechanism outside the house strains out any solids from the grey water and reduces the incidence of blockages in the irrigation system.

Diversion and treatment system

A more sophisticated method of grey water reuse might include a storage tank and biological treatment, a reedbed filtration system, a grease trap combined with a sand filter, or an aerated wastewater treatment system. One fairly simple and commonly available system connects to the discharge pipe and, when grey water enters the small tank supplied, a pump automatically sends the water to the required area via a flexible hose and turns off again when the container is empty. This and all similar systems are expensive to buy and install and it can be time-consuming to obtain the relevant permits.

Irrigation with grey water

All of the above systems require some means of getting the water to the garden and of dispersing it. This really needs to be considered at an early stage of your garden planning because surface application of grey water is not recommended, so digging trenches and installing pipes is involved.

One dispersal method is via absorption trenches fitted with slotted flexible pipe

(agricultural or agi pipe). Trenches are dug where desired, sloping away from buildings, and pipes are laid in them. The trenches are filled in with gravel or coarse mulch such as wood chips. Water seeps through the pipe and is dispersed through the gravel or mulch to nearby plants. An alternative grey water irrigation method, which is best used with filtered water to avoid blockages, is to use underground dripper lines.

A regulatory nightmare

One of the biggest deterrents to grey water use has been the difficulty encountered in trying to obtain permits from relevant authorities and, indeed, in finding out just what the regulations are for any given location. Every state has a different regulatory framework; councils and water authorities might have their own requirements; regulations for sewered and unsewered areas are different; storage, dispersal and type of grey water allowable all are regulated. No wonder a 2004 survey of grey water users found that 88% had not contacted their local council before installing a system.

Common to every state is the requirement for changes to the permanent plumbing of the residence to be carried out by a licensed plumber. In fact, the more sophisticated systems are too difficult for anyone without experience to install correctly. Another common requirement is that to divert pipes away from the existing system in sewered areas requires permission from the relevant water authority. Tasmania and Queensland do not allow grey water use in sewered areas.

Schematics of grey water use.

Grey water guidelines

If you have decided to pick your way through the regulatory maze and install a grey water system, there are some human, environmental, and garden-health matters to keep in mind.

- All grey water should be contained on your own property. There should be no runoff to gutters or adjacent blocks or seepage to groundwater.
- Do not use laundry water containing faecal matter.
- Do not store grey water for more than 24 hours without treatment.
- Do not use kitchen water.
- Be aware of what cleaning products you use. Many contain high levels of nutrients, especially nitrogen and phosphorus, unsuited to native plants.
- Phosphorus and nitrogen seepage and runoff into waterways is a major cause of toxic blue-green algal blooms.
- Alternate grey water and clean water use as much as you can to prevent build-up of undesirable salts and overly high levels of nutrients.
- Sodium from soap powders can damage plants and degrade the soil. Liquid detergents usually contain less sodium than powders. There are a couple of useful websites with details of detergent ingredients and what to look for:

www.lanfaxlabs.com.au
www.greenplumbers.com

- Many soap powders are highly alkaline and will tend to raise the pH of the soil.
- Do not use grey water on leafy vegetables or any eaten raw, and never sprinkle vegetables with grey water.
- Switch off the grey water diverter in wet weather when the soil is already saturated and runoff is likely to occur.

That's it. You're done. Garden planted, mulched and watered. Looks great doesn't it? Of course, you'll want to play around with your creation, adding and moving plants, installing water features and garden art, but the hard work, and the main expense, is over. Of course, there will be ongoing maintenance, with a major concern being pest control. Some ideas on how to achieve this the natural way are the subject of the next chapter.

SAFE SOLUTIONS
TO PEST PROBLEMS

No gardener wants their plants to be eaten by voracious pests and one of the most common questions about organic gardening is how this is to be prevented without recourse to a range of poisonous sprays promising a quick fix to all our gardening problems.

The answer to this question is, on one level, a very simple one and involves the development and encouragement of natural systems in which pests are controlled by predators. However, it would be a gross oversimplification to say that this provided the whole answer. We need to take a holistic approach, beginning with soil health and considering all the other aspects discussed in this chapter. None of them by themselves will control pests, but taken together they will result in a thriving ecogarden in which the 'pests' too have their role.

Cardinal rules

It is important to keep in mind my two cardinal rules. Rule one is that unless you know an insect, or beetle, or whatever, to be a pest, leave it alone – it is more than likely to be beneficial. Rule two is that even if you know it to be a pest, leave it alone unless it is causing significant damage – it too is likely to

be preyed upon by some other creature along the food chain if you just give the natural systems time to work.

Every rule has its exceptions. The exceptions to these rules are in the case of a newly planted area in which young plants are very vulnerable, or in the early stages of development of your ecogarden when the natural control systems are not operating. In these cases, use any of the methods outlined below as soon as pest damage becomes apparent.

Soil

Plants growing in healthy soil full of the necessary nutrients are more resistant to pest attack than are plants struggling to survive in poor soil. Good soil is the foundation of your garden and the starting point for an organic pest management program. Enrich your soil to be sure it contains all the required nutrients and trace elements, as well as a diverse active soil biota. Do this by using mulch, compost and organic fertilisers such as decomposed manures.

Test the pH if you think this could be a problem and take the relevant measures to correct it if necessary. Remember that most plants prefer slightly acid soil, so don't think you have to aim for a neutral pH. Add lime, dolomite or wood ash gradually and in moderation, as alkaline soil is more difficult to correct than acid soil.

Biodiversity

I discussed the importance of this in Chapter 2. Your flowering plants will attract birds, bees, wasps and other predatory and parasitic insects. Native shrubs and other plants with many small flowers over a long period of time will provide nectar and pollen for the adult stages of many of these beneficial insects.

Wasps

Nectar is important for parasitic wasps. These gardeners' friends feed on nectar in their adult stage. The larval stage, however, feeds on other insects and spiders. Some species of wasps lay their eggs directly on a host insect, perhaps a caterpillar, grasshopper or sawfly. When they emerge, the developing wasp larvae proceed to eat their host.

Other wasps build cells of mud. Into each cell the industrious wasp places a spider, caterpillar or grasshopper she has paralysed by injecting with a poison. She then proceeds to lay an egg in each, so each wasp larva is thus provided with its own pantry.

If numerous species of wasps abound in your garden, they will help control a multitude of pests for you, including, as well as those already mentioned, aphids, scale insects, lerps, moths, flies and beetles, for the small price of some nectar-producing plants and a favourable habitat. On the other hand, if a spraycan comes out every time a wasp is sighted, your garden will be the loser.

Bee flies

The female bee fly (a large hovering fly that produces a low hum) can lay up to 10 000 eggs in a knothole or crack in a dead tree or old fence post. The tiny larvae float on the wind until they land upon a host, often a wingless grasshopper or plague locust. They enter between the body segments of the host and develop by feasting on its soft body, which has a subsequently shorter life in which to damage your garden or a farmer's crops. Adult bee flies are also nectar feeders.

Aphids, hover flies, ladybirds and lacewings

In spring it is common to find aphids crawling over the growing tips and flower buds of your roses. Before rushing to the garden shed or the nearest garden shop for a poisonous spray, consider that you have some natural control measures close at hand. The gold and black hover fly, another adult insect that feeds on nectar and appears to particularly favour that of flat open flowers such as daisies and calendulas, has a larval stage as a maggot that feeds voraciously on aphids, often eating thousands of them.

Several species of the familiar ladybird beetles also feed on aphids, in both their larval and adult stages. Delicate green lacewings have a larval stage that feeds on lerps, scale insects and aphids.

Thrips

The thrip is another tiny insect much loathed by gardeners. However, as well as the pest species there are harmless and predatory thrips that help control other pests, including scale insects. Thrips fall prey to lacewing larvae, flower bugs and predatory and parasitic wasps.

Spiders

A healthy garden will be home to many species of spiders. Their webs are often masterpieces of architecture that capture a variety of pest insects to feed the spider's voracious appetite. They will also capture beneficial insects, but there is no need to go to the length of rising at dawn to free the bees that might have been caught in the webs around your house, as I have read of one person doing. Before brushing down those untidy looking webs from under the eaves, consider the benefit to your garden of this free pest control. Leave the webs for long enough and you might also be amazed to see small birds pecking the abandoned strands off walls to use in building their nests. A single spider can consume about 2000 insects in an 18-month lifespan.

Ants

These insignificant-looking little creatures play a vital role in a healthy ecosystem. The excavation activities involved in the underground nest making of soil-dwelling species aerate the soil and recycle nutrients from deeper levels to the surface. They scavenge the bodies of dead insects and other garden fauna, prey on a variety of pests, including crusader bugs and some caterpillar species, and are themselves prey for other insects. It has been said that the health of an ecosystem is directly related to the number of ant species present and that if ants were to disappear major extinctions of other species and partial collapse of some ecosystems would follow.

Insecticide sprays

Using a pesticide to destroy a 'pest' insect will also destroy a host of beneficial ones, thus reducing the biodiversity of the garden, as well as contributing to insect resistance and the never-ending treadmill of stronger, 'better' insecticides constantly being produced. As well, if we kill the pests (aphids for example) as soon as they appear, there will be nothing for the beneficial insects (ladybirds, bee flies, lacewings) to live on. They will depart to a more congenial neighbouring garden and the sprayer will be left breathing chemical fumes.

There are many more environmentally friendly sprays, including an ever-widening range of commercially produced products, available from nurseries and garden supply centres. In addition, it is possible to make up numerous sprays from common kitchen and garden ingredients. Even these less damaging sprays, however, must be used discriminatingly as many are nonspecific and will also kill or repel friendly insects. If you actively encourage diversity in your garden, whenever pests appear there will be predators following close behind.

Lizards and frogs

Lizards and frogs are efficient pest controllers and will more than repay your thoughtfulness in providing them with rocks and logs to hide under and ponds to breed in. A partially buried hollow log or length of wide-diameter plumbing pipe in a sheltered spot makes an ideal lizard hideaway.

Small lizards forage in the leaf litter and around rocks and logs for ants, springtails, beetles, termites and insect eggs. The larger species will eat worms, mice, grasshoppers, beetles, snails and smaller lizards. In the bush, lizards need fallen rocks and logs for shelter, as well as the leaf litter on the forest floor which harbours their prey, so try to duplicate these requirements somewhere in the garden. Around the vegie garden, lizards will consume slugs and snails, according to their size.

Frogs eat mosquitoes, small slugs and snails and other pests, consuming thousands over their lifespan. They need a dam or pond to breed in and the damp areas around them to hide in during the day. The pond should have a shallow area, needed for mating and egg laying. There should be plenty of shrubbery in which frogs can shelter, especially away from strong winds, which can quickly desiccate them.

For much of the year frogs hibernate under rocks and logs in moist shady spots, so you will need to re-create a similar habitat for them in your garden. Please don't remove rocks and logs from the bush, as they are needed there by the native fauna. If you have trouble finding natural-looking frog shelter, improvise an alternative by upturning some plant pots or half-burying a length of plumbing pipe. Never remove frogs from the wild as you could be unknowingly spreading diseased populations. Provide the habitat and the frogs will find you.

Birds

Bird numbers throughout Australia are in serious decline, with many species under threat of extinction. Creating a bird-friendly habitat in your garden will serve the dual purpose of assisting in bird conservation and controlling many insect pests. Plant flowering shrubs to provide nectar for honeyeaters, which also eat many insect pests. Leave 'untidy' areas for birds to nest in. Erect nest boxes in suitable trees, away from predators, if your area lacks natural nesting sites. If you have trees, alive or dead, with nesting hollows, leave them where they are. Nesting hollows are needed by many native bird species, and by other native fauna. Find out from your local council, environment group or Birds Australia if there are endangered species in your area and what you can do to help preserve them.

In return, the birds you attract will consume vast numbers of insect pests. Yes, they will also eat many friendly predatory insects, but on balance, they are more than worth the small effort needed to attract them and give them a home.

Thousands of interactions

I have mentioned just a few of the specific biodiversity relationships that can benefit your garden. There are literally thousands of these interactions going on in any vibrant garden ecosystem. With a little encouragement from you, the gardener, the life cycles of these myriad creatures will enhance your garden's health.

Plant choice

There are numerous small flowering plants that attract birds and beneficial insects, including alyssum, erigeron, marigolds, daisies, cosmos, grevilleas and other native shrubs. See Chapter 2 for many more suggestions. Wattles are an integral part of the landscape across much of Australia, providing nectar for a variety of birds and insects. Some species have nectaries at the

WATTLES FOR SMALL GARDENS

Acacia acinacea	Acacia amoena
Acacia buxifolia	Acacia conferta
Acacia drewiana	Acacia drummondii
Acacia farinose	Acacia flexifolia
Acacia hakeoides	Acacia jonesii
Acacia lanigera	Acacia notabilis
Acacia pubescens	Acacia strigosa

base of the leaves that produce nectar when the plant is not in flower. Do, however, choose species for your garden carefully as many will grow far too large for a small area.

The herbs parsley – fantastic, insects love the flowers – dill, tansy, borage, lemon balm, fennel and Queen Anne's lace also have insect-attracting flowers. Queen Anne's lace is often highly recommended for attracting insects, but it has the potential to become an environmental weed in some areas and I have found parsley to be much more effective. Perhaps the effectiveness of any particular flower in attracting insects depends on what else is in bloom nearby.

Keep in mind when choosing plant species that those suited to the climate and soil of your area are likely to grow more quickly, need less maintenance and be less stressed during extreme weather conditions, and thus are less likely to succumb to insect attacks or fungal outbreaks. I am not saying you can't grow tropical plants in Tasmania, for example, but you would have to work hard at duplicating as far as possible the natural growing conditions, and the plants would certainly be more susceptible to a range of problems.

Herbal leys

Planting a herbal ley is an interesting strategy that people with a large enough garden and a few fruit trees might like to try. Commercial organic orchardists are increasingly using herbal leys as a cover crop between fruit trees. This consists of a mixture of flowering herbs – tansy, yarrow, alyssum, borage, parsley, lovage, dill and marigolds, for example. These species attract predatory insects. Clover and lucerne are often included for their nitrogen-fixing and green manure benefits. Some orchardists also include a mixture of grasses such as Timothy, cocksfoot, prairie grass, tall fescue and polaris, but this is probably not necessary in a home garden situation.

The plants are allowed to flower and set seed, then the ley is mowed a section at a time so predatory insects always have an area to move to. The mowing returns nutrients to the soil by way of mulch. Plants reseed naturally, thus the ley is self-sustaining once established.

The system is dependent on there being enough water to maintain the ley as well as the fruit trees, such as the amount supplied by a sprinkler system. Where dripper systems are used, the herbal ley could be modified to incorporate a herb or two near each dripper outlet.

If you have an area of a few fruit trees planted in a built-up bed, a herbal ley could be incorporated in the bed around the trees. You would thereby be enriching the soil around the trees, reducing the amount of water needed

A modified herbal ley of borage, calendula and nasturtiums at the base of my four-grafted fruit trees.

and encouraging predator populations to keep orchard pests under control.

Companion planting

Diversity of plant species has another facet in your pest management program than that of supplying food and habitat for predators and parasites. Plants themselves have a web of complex relationships with each other. Consider the following ideas.

- Light and shade: shade-loving plants will be happy growing between or under the canopies of taller species.
- Leaf drop of taller species provides beneficial mulch for those below.
- Root exudates of one plant might be beneficial, or otherwise, to its neighbours.
- Species with a strong fragrance (many of the herbs, for example) can repel pest insects from other plants nearby.
- Ground cover or prostrate plants provide mulch and a cooler, damper root zone for their neighbours.

- Plants that fix nitrogen in nodules on their roots (legumes such as beans and peas, and even wattle trees) assist nitrogen-using plants grown nearby or subsequently.
- Many flowering plants attract beneficial predators that indirectly assist other species growing nearby by reducing pest numbers.
- Some plants just seem to like growing near each other while others have an antipathy for each other. Sometimes we know the reasons, often we don't.

Companion planting aims to make use of these complex relationships and can be especially beneficial in the vegie garden, where herbs in particular are used to assist in attracting predators and pollinators, repelling pests, enriching the compost or providing valuable mulch. Everywhere in the garden, however, variety encourages good garden health. Areas containing a single, or very few, species are an open invitation to pests, which, with no controlling factors, will thrive and destroy.

Companion planting lore

For centuries gardeners have observed and made use of plant relationships, and a recent renewal of interest in this ancient lore has led to the publication of many excellent books on the topic. If you want to explore this fascinating subject further, there is no shortage of reference material available. The following suggestions will help get you started. Some of them are not directly related to pest management, but have an indirect effect by assisting plant health.

- Plant aromatic herbs throughout the garden for culinary and medicinal purposes, to repel pests, to attract predators and pollinators when in bloom, to add to compost and mulch, to enjoy their fragrance, to make potpourri and for general garden health.
- Chamomile is known as the garden doctor and has long been used to improve the health of nearby plants. It will spread and act as a living mulch.
- Sage, rosemary, mint, marjoram, thyme, onion and garlic help repel cabbage moth.
- Some marigold varieties (*Tagetes* spp.) exude a substance from their roots that repels soil pests, particularly nematodes.
- Onion and garlic planted near roses, or their leaves used as mulch for roses, can help control black spot and add to the vigour of the plant.
- Roses surrounded by golden marjoram might flower more profusely and maintain better health.
- Hollyhocks planted near beans will attract bean flies and stop them from attacking the beans.
- In the vegie garden, interplant beans and leaf crops, or follow beans by a leaf crop, to take advantage of the nitrogen provided by the beans.
- Alternate rows of onions between the carrots are said to repel carrot fly.
- Surround broccoli, cabbages, cauliflowers and Brussels sprouts with onions (perennial tree onions are good) to repel cabbage moth and butterfly.
- Pyrethrum daisies make an attractive border and help repel a range of pests from vegetables. The flowers can also be

made into an effective spray. There is disagreement about the plant's effectiveness unless used as a spray, and probably by themselves they might not have much effect, but I've found them to be reasonably good as part of a holistic approach, even if only by way of breaking up a bed and confusing the pests.

- Yarrow has a beneficial effect on most other plants and spreads readily as a mulch if watered.
- Use fragrant herbs such as lavender, basil and rosemary to repel flies from around your doorways or outdoor living areas and to confuse insect pests in the vegie garden. If grown in pots, they can be moved to wherever you need them, including into the vegie garden.
- Corn, pumpkin and beans are a traditional trio grown together in the vegie garden. The beans, any climbing variety, planted when the corn is about 30 cm high, grow up the corn stalks, and the pumpkins spread and help maintain a cool, damp soil climate, which the corn appreciates.
- Pumpkins and potatoes do not like each other and will not thrive if planted together.

Not all combinations work in all areas. Try anything that sounds interesting to you and learn by observing the results. Remember – diversity is healthy.

Crop rotation

This aspect of preventing pest problems is particularly relevant in the vegie garden. If the same plant, or plant family, is grown in the same spot season after season, not only

A CONFUSION OF MARIGOLDS

The marigolds often used as companion plants to deter harmful soil nematodes belong to the Tagetes genus, commonly known as either French marigold (*Tagetes patula*) or African marigold (*Tagetes erecta*). The roots of these species exude a substance fatal to soil nematodes. *Tagetes patula* grows to 30 cm high and the same wide. *Tagetes erecta* is larger, growing to a metre or more high and 50 cm wide, although dwarf types are available. These and other Tagetes species all have a pungent aroma that can confuse pest insects. Grown among vegetables they repel white fly, carrot fly, pumpkin beetles and a range of other pests, and confuse the cabbage white butterfly.

The other plant commonly called marigold is a different genus, *Calendula officinalis*, often referred to as pot marigold or English marigold. This is an edible herb, used in salads and cooked dishes. In the Middle Ages it was regarded as a remedy for a variety of ills from indigestion to smallpox, and was used in stews as a cheap substitute for saffron, hence the name pot marigold. Calendula can be used as a companion plant to attract beneficial parasitic wasps.

could soil-borne disease organisms build up, but also the soil is constantly being depleted of the same nutrients. As different plants have distinct nutrient requirements and generally are subject to different pests and diseases, many potential problems can be prevented or minimised by varying the crop grown in any area each season.

One recommended rotation is the solanum family (tomato, potato, eggplant, capsicum, pepper), followed by brassicas (cabbage, cauliflower, Brussels sprouts), then legumes (all peas and beans), then everything else. Another is legumes, the onion family (onions, garlic, chives, leeks, shallots), brassicas, then everything else.

In my own built-up vegie beds, I don't usually follow a strict rotation because each bed is usually a hodgepodge of interplanting. I do avoid growing members of the solanum family in the same bed for two or three years, and either precede or follow the nitrogen-hungry brassicas by legumes, which convert atmospheric nitrogen in their root nodules to a form plants can use in the soil. Often, however, I simply interplant these two families. I avoid having pumpkin or squash in the same bed as potatoes, as neither will thrive.

If you grow vegies and can't rely on your memory from one season to the next, it is a good idea to keep track of crop rotations in a gardening diary.

Hand-picking/crushing/jet of water

Don't underestimate the value of the simple pluck and destroy, or crush between gloved fingers methods. These can be especially appropriate if you have a small garden, and older children can prove to be enthusiastic allies in this job. Do ensure it is only the real pests you, or they, are destroying. A moderately strong jet of water from the hose is also effective in dislodging many pests.

Christmas beetles and sawfly larvae can easily be plucked off young affected plants and either fed to poultry or crushed underfoot. Young family members often enjoy catching grasshoppers in a net when they are in large numbers. Slugs and snails can be squashed underfoot on a damp night or early morning, if you are not too squeamish. Aphids on the roses can be crushed between gloved fingers or sprayed off with a hose. Don't be too thorough though; you need to leave some tucker for the ladybirds, hover flies and other beneficial insects.

Physical protection

You can use a variety of devices to protect young plants or vulnerable seedlings. I cut the bottoms off two-litre juice bottles and put them over individual seedlings in the vegie garden, pushing them a centimetre or two into the soil. This has the added advantage of creating a mini hothouse environment when early spring temperatures can be variable.

Juice bottles used to protect young seedlings in the vegie garden.

Shell grit around seedlings will keep off slugs and snails.

Later in spring and through summer, to avoid cooking seedlings, it is more appropriate to use large coffee or fruit tins with top and bottom removed and an onion bag or old pantyhose stretched over the top. Remove the bottle or can when the seedling begins to outgrow it, by which time it will usually be able to withstand a little munching and still continue to thrive.

Bigger pests

If bigger pests such as possums, domestic pets, poultry or kangaroos are a problem, it is necessary to use appropriate fencing to keep them off the vegie garden, or wherever else you don't want them. Guards fashioned from wire netting are useful in many cases and are infinitely reusable. Lengths of wire netting placed over a garden bed will prevent birds from scratching or pecking at emerging seedlings, or domestic pets from using the bed as a toilet facility.

Slugs and snails

A band of wood ash, crushed eggshell, sawdust or shell grit (as fed to poultry)

around seedlings will keep slugs and snails off, as they dislike crawling over the sharp particles. Rice hulls can be used in a similar manner, as can hair, either human or animal. Make sure the barrier leaves no gaps for the gastropods to crawl through. Ash will need to be replaced after it rains, or after overhead sprinkling. The other barriers will last a little longer; with luck, for long enough to allow the seedlings to grow and harden past the most vulnerable stage.

Copper is said to deter slugs and snails. Apparently, crossing a copper barrier is akin to an electric shock for a gastropod. There are rolls of copper band commercially available to place around individual plants.

Birds and possums

Nets over your fruit trees are the most effective means of preventing birds and possums from eating your fruit in areas where they are a problem. There are different types of bags – mesh, cloth, waxed paper – you can buy to cover individual fruits with to keep off

These polypipe hoops are in place over an apple tree ready for netting to be attached.

possums, birds and fruit flies. These seem like an unnecessary amount of bother to me. Better, I think, to cover the whole tree, or selected branches, with netting.

Grasshoppers

When grasshoppers ate the leaves off some young fruit trees, and then began to eat the bark, I wrapped and tied some green mosquito netting around each tree so the leaves had a chance to grow back and the trees to recover.

Traps and lures

Some pests can be attracted to a trap. They are either killed by the bait in the trap, usually by drowning, or are subsequently destroyed. Carnivorous plants can be included in the category of traps, but are indiscriminate, with no guarantees they will only kill pests. However, they are worth a try as part of a holistic approach, and could be beneficial grown in pots placed around the

entrances to your home to trap flies, or placed in European wasp flight paths.

Sticky traps

Different insects are attracted to different colours. Yellow is said to be attractive to white flies, aphids, leafhoppers and moths, and blue to thrips and leafminers. This knowledge can be put to use to make traps for the relevant insects. You can cover a piece of coloured plastic, perhaps cut from an ice cream container, with a sticky substance such as petroleum jelly and leave it near where the pest has been seen. The colour will lure the insects and they will be stuck to the sticky surface. Commercial products are available.

Slugs and snails

Grapefruit or orange skins, potato shells, shredded carrot, cabbage and lettuce leaves, banana skins, bran and yeast have all been used to attract slugs and snails. Place your chosen bait in a damp, dark corner and cover it with a sack. Check the area every day and dispose of the pests. Try a variety of baits, as results are variable according to place, climatic conditions and the species of the pest.

Alcohol is attractive to slugs and snails. Half-fill margarine containers with a 50:50 mixture of beer and water, and put them in a sheltered spot in the vicinity of slug/snail activity. Milk used in a similar way is said to be effective, as is Vegemite or Marmite dissolved in water.

A deadly bait can be made from pyrethrum powder mixed with molasses and enough water to form the mixture into small balls or pellets. Place them where slugs/snails are active, but ensure that pets and small children cannot get to them – inside a length of polypipe is a good idea. Collect and dispose of the bodies daily.

In recent years a couple of different brands of commercial products have been released that are safe to use in organic gardens. They are claimed not to harm pets or birdlife. The active ingredient is an iron complex that, once eaten by the pest, replaces the copper in its blood that is necessary to transport oxygen (in humans this task is dependent on iron). Uneaten pellet residues simply break down to add a trace of iron to the soil.

Earwigs

These gather under wood or bark and can be a problem if present in large numbers. They seem to be very fond of unripe fruit, especially nectarines, and will chew all around the skin. Place pieces of corrugated cardboard in problem areas. Check every few days, dispose of earwigs and replace cardboard until no pests are found. Another method is to place crumpled newspaper inside an upturned plant pot. Again, check and dispose of pests every few days.

Fruit flies

A very popular method of fruit fly control with organic gardeners and orchardists is the use of lures and traps. A mixture, which attracts the insect, is placed inside a trap. The pest is then killed either by the toxicity of the lure – pyrethrum, for example – or by drowning in the mixture. Some lures are said to be specific to male fruit flies and some to females, while some are general. It is common for a homemade lure to consist of wet, sweet and yeasty ingredients.

Male lures include Vegemite and citronella oil. Female lures might consist of orange pulp or rotting peach or plum. All should be mixed with water. To ensure pests are killed, some pyrethrum can be added to the mixture.

In addition, there are several organically acceptable commercial products. Some of these use pheromones to attract male fruit flies and one is a formula based on essential oils of native plants that attracts and kills the male flies.

Any lure needs to be placed in a trap. Traps are commercially available, but it is quite simple to make your own. All you need is a bottle, jar or tin in which the lure can be placed. Make a small hole in the container to allow the flies to enter. Hang several or many, depending on severity of the problem, in and around your fruit trees. Check frequently and remove any old lure and dead flies.

Safer sprays

No matter how successfully you have created an eco-friendly garden of teeming biodiversity, and incorporated the other pest management techniques mentioned in this chapter, there will be times when pest numbers are at unacceptable levels. It could be that your garden is still in the establishment stage and not yet supporting a wide diversity of predators. However, it can also occur because of extreme weather conditions, plagues of a pest over a wide area, or simply because it is early in the season and the pests (aphids, for example) are active before predators (ladybirds, hover flies, lacewings, wasps) have built up enough to control them.

These situations are usually short-lived, particularly in the latter example, and will correct themselves without any action on your part. However, if you decide there is a need to take action to prevent excessive damage, there is a plethora of homemade sprays and many organically acceptable commercial products to repel or kill most pests.

Keep in mind that even the mildest spray will have an effect on non-targeted species as well as pests. Most homemade sprays, in common with the commercial insecticides, albeit less toxic, are nonspecific, so you'll be destroying or repelling beneficial predators as well. Some sprays, while being harmless to mammals, can affect fish and other aquatic life, and some (tomato and rhubarb sprays, for example) are harmful to humans, so due care must always be taken in their manufacture, use and storage. Keep containers and equipment specifically for this purpose, label clearly any sprays that you store and keep them out of reach of children. I seldom use any of the homemade sprays, having a strong disinclination for the fiddling involved and preferring to let the predators sort out the pests wherever possible.

There are dozens, if not hundreds, of possible sprays. Following are just a few examples of those made from common ingredients that most people will have available and some suggestions about possible commercial products. I have avoided using any such

ingredients as flour that will clog up your spray nozzle. Be sure to strain out any solid ingredients carefully as a blocked spray nozzle is hard to clear and guaranteed to incite temper tantrums.

All the homemade recipes break down quickly and will need to be re-applied every few days if pests are still prevalent. If you find yourself spraying frequently, you will need to aim in the long term to increase the biodiversity in your garden. Always be sure to wash any sprayed fruit and vegetables well before using them.

Garlic spray (1)

Garlic spray is a general-purpose insecticide, effective against a range of pests, including snails, aphids, codlin moth, white butterfly, caterpillars and wireworms. Commercial sprays are available. Soak 90 g chopped garlic in 2 teaspoons vegetable oil for 48 hours. Dissolve 30 g grated pure soap in 600 mL warm water. Add to garlic. Filter well and store in a sealed plastic or glass container. Dilute using 1 part solution to 99 parts water to begin with, and strengthen if necessary.

Eucalyptus spray (2)

General purpose for most soft-bodied pests, including caterpillars, also cockroaches, ants and beetles. Use a 2% solution of eucalyptus oil in water with a little detergent or grated pure soap as an emulsifier and wetting agent.

Molasses spray (3)

Will deter beetles, nematodes and caterpillars. Molasses contains a high level of potassium, which could be beneficial to plants. Dilute a tablespoon of molasses in 4 litres of hot water. Leave to cool. Use on fruit trees, vines and vegies.

To control nematodes, apply the solution as a soil drench to affected areas with a watering can. It does also affect earthworms, so use with care and only if absolutely necessary.

Solanum spray (4)

(tomato, potato, capsicum)

General purpose, especially for chewing pests. Chop 250 g of the leaves. Simmer for 30 minutes in 4 litres of water. Use an old saucepan kept especially for this purpose, as this spray is one of the most toxic in the homemade repertoire. Strain. Dissolve 25 g grated pure soap or soap flakes in the mixture while still warm. Leave to cool. Dilute 1 part solution to 4 parts water.

Rhubarb leaf spray (5)

Use for aphids. Chop 250 g rhubarb leaves and place in a large saucepan, kept especially for the purpose, with 600 mL water. Bring to the boil and simmer for 30 minutes. Strain and cool. Add a handful of soap flakes dissolved in 300 mL warm water.

Tomato spray (6)

Effective against caterpillars of all types (including benign or beneficial, so take care when using and target the pest only) and grasshoppers. In an old saucepan kept for the purpose, boil 2 cups chopped tomato leaves and stems in 2 cups water. Strain and cool. Dilute with a further 2 cups water before use.

Chilli spray (7)

For caterpillars and red spider mites. Blend 2 cups chillies with 2 cups water. Strain. Use immediately.

Feverfew or pyrethrum spray (8)

Both feverfew and pyrethrum make a general-purpose spray that will kill most pests, but will also kill predators. The following quantities are suggested for pyrethrum; double the strength if using feverfew. Strength will vary with different cultivars, so start with a weak solution and strengthen if necessary. Do not boil, as the fumes are toxic. Soak 2 tablespoons of the flowers in a litre of hot water for an hour or two. Strain before use.

There are commercially available pyrethrum-based sprays, which might consist of pyrethrum only or be mixtures of pyrethrum, garlic, chilli and other natural ingredients. Pyrethrum will break down in sunlight in anything from two to 48 hours and has a toxicity period of 12 hours. If sprayed at night, it is less likely to affect bees and other beneficial insects.

Wormwood spray (9)

Use for soft-bodied pests such as caterpillars and aphids. Place 500 g leaves in a metal bucket. Cover with boiling water. Soak until cool. Strain. Use undiluted.

Syrup or treacle spray (10)

This is said to effectively repel grasshoppers and locusts, which avoid the sticky surface. Mix 2 cups syrup or treacle in 2 to 4 litres (experiment to determine effective strength) water. Spray affected plants and garden perimeters. Adding treacle to the tomato spray (6) will help it stick to the targeted area and increase its effectiveness.

Oil spray (11)

Use this for most sap suckers, including scale, thrip, mealy bug, white fly, spider mite, and other over-wintering pests and their eggs and larvae. The oil suffocates the pests and/or their eggs. Use a vegetable oil after leaf drop and before new buds burst. Preferably use in temperatures below the mid-twenties or leaf damage might occur.

Grate 250 g pure soap (or use flakes). Add to 2 litres of oil in a large old saucepan. Boil and stir until soap dissolves. Use about a cup of oil/soap to 20 cups of water. Dilute only the amount you intend to use at once otherwise the remainder will separate and be useless.

Alternatively, mix together 125 mL fish oil, 125 mL vegetable oil and 1 cup detergent. Dilute a third of this mixture with 4.5 litres water.

There are various commercial oil sprays; at least one contains eucalyptus and tea tree oils. They are suggested for citrus leaf miner, aphids, mealy bugs, white flies, mites and scales.

Onion spray (12)

For most sap-sucking pests, including aphids, mites, scales and thrips. Chop onions and cover with boiling water. Leave in a sealed container for one or two days. Strain. Dilute with an equal quantity of water and spray affected plants every few days until pests have gone.

Herbal repellent spray (13)

This general-purpose mixture will repel pests rather than kill them. Blend a mixture of any of the herbs known to be insect repellents – garlic, onion, mints, lavender, sage, rosemary, basil – with enough water to make a slurry. Store a day or two in a sealed container. Strain. Add an equal quantity of warm water in which you have dissolved a handful of grated soap, or flakes, to aid adherence.

Bug juice (14)

This is not for the squeamish, but does have the advantage of being specific to the targeted pest and is often recommended against grasshoppers. First, catch your pest, about a cupful. Cover with 3 cups of water. Some recipes say to blend to a slurry; others say to just soak for 24 hours. Strain well. Dilute with 50 parts water. If this doesn't work, add more 'juice' to make a stronger concentration.

Milk spray (15)

This is reputed to be effective against red-legged earth mites, which dislike the waxy layer left by the milk. Make a mixture of 1 part milk to 9 parts water and spray it on vulnerable seedlings. Repeat every few days until seedlings are hardened enough to withstand the pest, or until the pest is no longer evident. This spray, particularly with the addition of a small amount of honey dissolved in warm water, is reputed to attract large numbers of beneficial insects and to act as a tonic for any stressed plants.

Quassia spray (16)

Quassia is a South American tree, the chipped wood of which can be made into a spray to discourage possums from eating your fruit. It makes the fruit taste bitter and unattractive to the possum, and also to you, so wash sprayed produce well before eating it. It can also be used to control aphids. Soak a teaspoon of quassia chips overnight in 4 litres of water. Strain and spray. Re-apply every two weeks until the marauders get the message.

Dipel (17)

Dipel is a commercial product containing bacteria, *Bacillus thuringiensis*, which is a stomach poison to caterpillars. A spray is made up and applied every three to five days to control the chewing pests. It breaks down in sunlight, so re-application might be needed.

Fungal diseases

Fungal problems are more likely to proliferate in very acid soils, as fungi (particularly the undesirable ones) continue to flourish at low pH levels, whereas general soil biota prefer a pH of 6 to 7. This means there will be more of the fungi at lower pH levels and fewer other organisms to control them.

Fungal problems such as curly leaf, black spot and powdery mildew are less likely to occur in a slightly acid to neutral pH and in soils containing plenty of organic matter and teeming with soil biota. So, improving the soil and encouraging soil biota are the first and ongoing priorities in prevention of fungal diseases. Another consideration is growing plants that suit the area.

There are various organically acceptable sprays that are effective. They are not once-off quick fixes, but need to be repeated on a

SPRAYS FOR COMMON PESTS AT A GLANCE

Pest	Spray number
Ants	2, 3, 4, 8, 10
Aphids	1, 2, 5, 8, 9, 12, 13, 16, 28
Beetles	3, 6, 8, 14
Borers	8
Caterpillars	2, 3, 4, 6, 7, 8, 9, 13, 17
Cockroaches	2, 4, 8
Codlin moths	1, 8, 13
Grasshoppers and locusts	7, 8, 10
Mealy bugs	1, 11, 12, 13
Mites, red spider and bean spider	1, 11, 12, 13, 15, 16, 28
Nematodes	3
Possums	16
Scales	1, 8, 11, 12, 13, 24, 25, 28
Snails and slugs	1, 8, 13, 22
Thrip	1, 11, 12, 13, 24, 25
White butterflies	1, 8, 13
Whiteflies	1, 7, 9, 12, 13
Wireworms	1, 8, 13

regular basis, while at the same time you are improving the soil. Some of the commercial antifungal sprays are also effective insecticides; lime sulphur, for example, can be used to control a range of fungal diseases as well as scales, mites and aphids.

Seaweed spray (18)

As well as being antifungal, seaweed spray contains a range of trace elements and is a great plant tonic. It is helpful for treating curly leaf, black spot, brown rot, canker, silver leaf and powdery mildew, but will not on its own eradicate them. It is better as a preventative and all-round plant tonic. There are many commercial products available.

Removing seaweed from the shore to make your own spray could have adverse environmental consequences and might be against your council's bylaws. If you do legally have access to seaweed and can remove small amounts without adversely affecting the environment, it is quite easy to make. Collect a bucketful of seaweed and wash the salt off it. Cover with water and leave to soak for two to three weeks. Strain and dilute with enough water to make it the colour of weak tea.

Bicarbonate of soda (19)

For powdery mildew on apple trees, cucumbers, marrows, roses, strawberries and grapes. Dissolve a level teaspoon of bicarbonate of soda in 2 litres of water.

Bordeaux spray (20)

Bordeaux is available commercially, or you can make your own. It is used for curly leaf, black spot and brown rot.

Dissolve 125 g copper sulphate in 2.25 litres of hot water in a plastic bucket. In another bucket mix 60 g hydrated (brickies') lime with 2.25 litres cold water. Strain each mixture through pantyhose into the spray container to combine them. Use straight away.

Another recipe is to dissolve 50 g of copper sulphate in 1.5 litres of hot water and 50 g of hydrated lime in 1.5 litres of cold water.

Combine the two and add enough water to make the quantity up to 5 litres.

Adjust quantities of both recipes to suit the size of your garden and the size of your sprayer.

Unless you are friendly with a brickie or an organic farmer who makes their own spray, you could have difficulty buying the hydrated lime in a small quantity – it is normally sold in large bags.

Use this spray on fruit trees in autumn just after the leaves fall, in mid-winter when the plant is dormant, and in late winter to early spring when the new flower buds are just showing pink – generally known as pink bud stage.

A Bordeaux paste can be used on trees affected by collar rot, fungi or wounds. Make up a paste with 60 g of hydrated lime and about 300 mL of water. Use straight away by brushing onto the affected area.

Chamomile tea spray (21)

Use this to control damping off in seedlings or mildew on grapes and squash. Soak 25 g of leaves in a litre of boiling water. Leave to cool. Strain. It can be used as a spray, but I use a tea made from chamomile tea bags to water newly planted seeds and seedlings in punnets.

Condy's crystals/potassium permanganate (22)

This is both insecticidal and fungicidal. It is used to control aphids, slugs and powdery mildew. Mix 10 g Condy's crystals, 3 litres water and 10 mL vegetable oil. Spray every 10 days while pests are active.

Compost spray (23)

This is a general fungicide said to be effective against grey mould, potato blight, clover rot, apple scab and even phytophthora (root rot). Use well-rotted compost (six months old is recommended) free of heavy metals. Soak compost in a drum in three to nine times its volume of water. Stir. Leave to soak for a week or longer. Skim off liquid, strain well and use undiluted.

After a week of soaking it is effective against phytophthora and after three weeks it is said to be strong enough to use for botrytis (grey mould) control. The compost spray doesn't kill the fungi, but prevents them from settling and colonising the plant. Spray up to once a week at humid times of year and repeat after rainfall. Repeated sprays are more effective.

Vegetable oils (24)

To control powdery mildew and sooty mould use sunflower, olive, canola, peanut, soybean, grapeseed or safflower oil. Mix 3 drops of liquid detergent, 10 g of oil and a litre of water. Use immediately.

White oil (25)

This is used mainly for sooty mould on citrus trees, which is caused by scale insects. White oil is a light petroleum-based mineral oil, which is accepted for use in commercial organics and available at any nursery. It is said to have a short residual life and minimal effect on beneficial insects. If treatment for sooty mould seems to be necessary, the vegetable oil spray above should be effective.

Milk spray (26)

I have seen a dilution of half milk and half water recommended as an effective treatment for mildews. You would really need to keep your own cow for this to be practical or economical. It would probably make the garden smell like sour milk as well. It is worth trying a weaker dilution, such as 1 part milk to 9 parts water, as suggested for red-legged earth mites.

Chive spray (27)

This is used to control mildew on squash and zucchini plants. Vitamise a cup of chopped chives in 3 cups of boiling water. Leave to cool. Strain and dilute with an equal quantity of water.

Lime sulphur spray (28)

This commercial spray is often used to control curly leaf, black spot and powdery mildew. The concentrated liquid is diluted in water at the

SPRAYS FOR COMMON FUNGAL PROBLEMS AT A GLANCE

Fungal problem	Spray number
Black spot	18, 20, 23, 28
Brown rot	18, 20, 23
Canker	18, 23
Curly leaf	18, 20, 23, 28
Damping off	21, 22
Powdery mildew	18, 19, 21, 23, 24, 26, 27, 28
Silver leaf	18, 23
Sooty mould	24, 25

rate recommended on the bottle before use. It should not be used in hot weather, over 32° Celsius, or within 10 days of using an oil spray.

Commercial products

There are commercially prepared insecticides and fungicides available in garden supply centres that are safe to use in organic gardens, some I've mentioned above. Look for them and read the labels before making your purchase; most will have a certified organic logo or the words 'chemical free'. For a catalogue of products related to organic gardening, including seeds, books, fertilisers and pest management products visit the website: www.greenharvest.com.au.

ACTION PLAN

- Continue developing healthy soil.
- Encourage biodiversity.
- Experiment with companion planting.
- Rotate crops.
- Physically remove and exclude pests.
- Use traps and lures.
- Try homemade and commercial organically acceptable sprays.

THE PLOTS DEVELOP

To see how everything comes together it is time to return to the development of my garden plots. We had numerous ideas when we started, some quite clear, others more vague. How did they translate into reality?

In general, it is wise to work on one plot at a time, according to your predetermined order of work. In my case, I had a load of pebbles to spread in several areas and as this was a physically demanding and time-consuming job, I could only do a little at a time. While this was underway, I worked on some plots where progress was easier and quicker.

Between house and western fence

Most houses have a similar narrow strip, usually dedicated to providing a home for the clothesline, hot water service, heating unit and rubbish bins. I was determined from the start that this utilities area was going to be an attractive part of the garden. In fact, it was probably the first plot that began to take shape in my mind after we'd seen a similar treatment of narrow space at a display home village. Being on the western side of the house, I thought it would be a hot area very suitable for a pebbled succulent garden.

For added privacy we extended the height of a section of the fence with plastic lattice, planning to grow creepers up it.

Preparation

A friend gave me some succulents from her garden before I was actually ready to plant anything – it was a drought summer and the ground was dry and hard. I scraped some shallow holes along a section beside the fence, planted the succulents and kept them watered by hand, intending to move them later. There was no preparation of the soil; nothing at all was added to this small patch. The succulents thrived so well on this minimalist treatment

Lattice extended the height of the fence and was planned to support some climbing plants.

The succulent garden a few years down the track after we had removed the palm trees and painted the fence.

that I decided they could stay where they were. When it finally rained and became a little cooler, development of the area could begin.

This had been a dumping spot for the builders, so there was a lot of rubble to be cleaned up before the grass and weeds could be slashed with a brush-cutter.

The narrow strip beside the fence where the succulents had already been planted was extended. I dug it, without turning the soil over, to aerate the soil and remove weeds, and then edged it with leftover bricks before raking a few barrow loads of topsoil over the dug portion.

The remainder of the area was covered, a few square metres at a time, with flattened removals boxes, well overlapped, except for a strip extending about 50 cm out from the house. This strip was covered in heavy-duty builders' plastic as a termite deterrent.

Mulch used

Sheet mulch of boxes, as above, was laid on top of brush-cut weeds. The cardboard was wet to make it stay in place and be less slippery for the next stage of wheeling the barrow onto it. A load of river pebbles, eight cubic metres, had been bought from a local quarry and dumped on what was to be the vegie garden. I tipped a wheelbarrow-load of pebbles at a time on top of the cardboard, and raked it evenly after every few loads.

Compost and sugar cane mulch was spread along the strip next to the fence after planting.

All this took weeks of work to do a little at a time.

Plants

The lightly dug strip adjacent to the fence was planted with *Clematis aristata,* a native creeper that would eventually grow up the lattice to provide shade and screen the house from the property next door. Some palm trees were intended to provide height and a tropical feel. Erigeron and heuchera at the base of the palms would spread to be a living ground cover mulch and their flowers would attract beneficial insects.

When it was time to plant succulents in the pebbled area, plants were positioned in their pots to determine their final position. Next, the pebbles were pushed aside and holes cut in the cardboard. Compost for nutrients and fine screenings – another leftover – for drainage were worked into each hole prior to planting.

Watering

Microsprays water the strip along the fence. The succulents in the pebbles are watered, very infrequently, with a hand-held hose. As a general rule of thumb, the more fleshy the succulent's leaves, the less frequently it needs to be watered.

Evaluation

Though it is a hot spot, it is not as hot as I expected due to shade from trees next door. These overhang the fence and need to be pruned back at times, but, on the other hand, they provide an attractive backdrop to my plantings. The succulents have thrived and need very little attention or water. Most of the clematis are growing well, but two died suddenly in the first summer, I think as a result of extreme heat reflecting from the light-coloured metal fence. The palms struggled, perhaps for the same reason, and were eventually removed and replaced with more succulents.

The cardboard under the pebbles decomposed more quickly than I had anticipated and some weeds grew among the pebbles and needed to be pulled out, a fiddly job, until I realised that some of them, oxalis in particular, were easy to kill by pouring boiling water over them. The onion grass was not so easily vanquished. In a few patches where it grew I pushed the pebbles aside, dug out the weeds (there is a bulbous root you need to get out) and re-covered the spots with thicker cardboard. Despite my wish to recycle this resource, the removals boxes were too thin for the job – commercial weed mat or thick white goods boxes would have been better.

This area was significantly redeveloped after the 2006–2007 drought. More succulents replaced the palms and erigerons and the fence behind was painted blue. I think it is a vast improvement and am very happy with it.

Nature strip

This is a town with a culture of nature strip gardens. Along our road there are no footpaths, but there are wide nature strips on which many residents have planted gardens of one sort or another. Hence we felt confident about extending the planting of our narrow frontage into the adjacent nature strip. Many councils have vetoes or restrictions on nature strip planting, so it is safer to check first if you are thinking along these lines. Do ensure that anything planted on the nature strip does not impede traffic or pedestrians.

Our aim was to have a privacy screen from the road and a buffer from the south wind that would also provide food and habitat for insects and native birds.

Preparation

Individual holes were dug and compost worked in. The mound, moat and mulch method was used here to facilitate watering.

Mulch used

The initial mulch consisted of leaves scraped from under the existing gum trees plus sugar cane mulch. This has since been topped up with used cat litter, more leaves and some straw.

Plants

Plants chosen for this spot needed to be tough to cope with competition from the existing mature trees and the buffeting of the south wind. I also wanted them to be fast growing and have insect- and bird-attracting flowers. They included a couple of indigenous

The nature strip soon after planting.

wattles, native hibiscuses, grevilleas, several small and two larger gum trees, bird of paradise plants and native grasses.

Watering

A hand-held hose is used to water plants individually. The mound, moat and mulch system enables deep watering that soaks in well without runoff. By the end of the second summer watering had been extended to two-week intervals. Most of these plants should not need supplementary watering after a few years.

Evaluation

Most plants are growing well and my aims are being realised. As early as the first summer wattle birds were sipping nectar from the grevilleas. The bird of paradise plants suffered from frosts and were replaced by more frost-resistant natives. The watering moats around the plants are gradually eroded by skirmishes with the mower and by blackbirds scratching around the plants, but it is a simple matter to pull out encroaching weeds and re-form the moats.

About a year later, the plants are growing and flowering well.

Driveway verges

The driveway is semicircular and gave me opportunities for two border plantings, one along a section of the western fence and the other on the eastern verge of the driveway.

Preparation

A narrow strip along the western edge was dug lightly, without turning the soil over, weeds were removed and topsoil and compost were added and raked over evenly. Planting holes had decomposed cow manure, compost

Planting and development of 'Orange Twists' along western driveway verge.

and a pelletised soil improver and mulch added. This bed was edged with red gum sleepers.

On the eastern side, we had some slabs of red gum trunk, inherited with the block, positioned at intervals by the contractor who formed the driveway. Around and behind these I built up some mulch and edged the back of the bed with leftover bricks. Compost and decomposed manure were added to the planting holes.

Mulch used

On the western side sugar cane mulch and a scattering of pelletised soil improver and mulch were used initially. Subsequently, leaves shed by the mature gum trees on the nature strip have continually built up in front of the red gum edging. From time to time I scoop these up and add them to the mulch, along with weeds pulled occasionally from the gravel driveway and leaves trimmed from the irises.

The eastern bed was built up with stable litter of urine- and manure-impregnated sawdust, compost and sugar cane mulch. It has subsequently been added to with gum-leaf detritus and weeds.

Plants

These plots needed tough plants to cope with both wind and radiant heat from the gravel driveway. I chose plants that would not be too spreading and end up scratching cars and people as they come in and out.

The western side along the fence consists of a border of 'Orange Twist' (*Syzygium australe*) with alternating white and apricot irises in between.

The eastern side had blue fescue grass between the red gum slabs with alternating dwarf 'Munstead' lavender and New Zealand flax, *Phormium* 'Yellow Wave', behind.

Watering

Seeper hose waters both these borders, for half an hour once a week.

Evaluation

Initially, the western border had some *Phormium* 'Yellow Wave' included among the irises. These died during the first summer, probably because of reflected heat from the fence. The 'Orange Twists' and irises are thriving and it won't be too long before the fence is entirely hidden by the 'Orange Twists'.

In the eastern border the 'Yellow Waves' did exceptionally well, for the first three years, but some of the lavenders struggled and there were some losses. During the 2006–2007 drought I decided to give these plants away to a friend with more water available than I had. When the drought broke I replaced them with cushion bushes and a hardier phormium.

Walk-through shrubbery/eastern driveway copse

A shrubbery area designed in meandering rows to be walked through and around abuts the eastern driveway border, which forms its outer row. More hardy fauna-attracting species, both native and non-native, were chosen for this plot, with the added aim of having something flowering at all times of the year to provide interest for the human residents, and nectar and pollen for the other creatures that live and visit here.

Preparation

Plants were laid out in their pots and left for several weeks while I played around with them to obtain a pleasing design. Individual holes

From start to finish: eastern driveway copse development (see also next page).

Flowering plants in the copse attract many pollinating insects.

were dug and compost, decomposed manure and a handful of soil improver and mulch worked in. Some spots were very compacted and needed a crowbar to break up the claypan. These were given an extra helping of compost and topsoil.

Mulch used

Initially, the mound, moat and mulch method was used, with each plant being mulched with sugar cane mulch. After a short time I decided to sheet mulch around the whole area so the mulch would form walkways, rather than leave grass paths that would need constant mowing. Thick boxes from fridges and televisions formed the base. These were covered with some of the shredded tree mixture we obtained from a local contractor.

Plants

Each row has a different mixture of plants, mainly chosen for hardiness and biodiversity reasons. They include Geraldton wax shrubs, sea hollies, irises, a little silver-leaved ground cover called snow-in-summer, buddleias, callistemons, a grevillea, hakeas, baeckias and a protea.

These shots show the development of the copse from different angles.

When the contractor formed the driveway and placed the red gum slabs along a section of it, he also left a section of trunk in this area of garden. I couldn't move it and wasn't going to pay for him to come back and do this small job. What could I do but incorporate the stump into the garden? I used the stump as a starting point for a more or less circular brick edging. Inside the edging is a bay tree at the centre surrounded by arctotis and bromeliads. The mulched walkways around the rows of plant all lead to this circular bed.

An existing adjacent red gum casts dappled shade over the whole area for some of the day and replenishes the mulch with dropped leaves and bark.

Watering

All four rows are watered by seeper hose (laid under the sheet mulch), in four separate lengths attached to a feeder length of polypipe, which snaps onto the normal hose. Each row has a separate tap and I usually water one at a time for about half an hour. The circular bed is watered by hand-held hose or sprinkler.

Evaluation

Everything is growing well. The buddleias will need constant trimming to ensure they don't overpower less vigorous species and make the walkways impassable. Many bees and butterflies visit the buddleias, sea holly and Geraldton wax flowers, but I removed the sea holly because it was showing a tendency to sucker and I really don't want its prickly flowers everywhere. The irises under the buddleias don't get enough sun and will need to be moved.

In the process of making a floodway into a feature.

Pond and creek-bed

When we bought the block this low-lying area was one of the features that interested us. We knew we could turn this 'problem' seasonal floodway into something stunning. Runoff from a poor town drainage system flows through here after heavy rain. We wanted to both improve the flow and intercept some of it into a small pond.

Preparation

A backhoe driver was hired to deepen what was naturally the lowest part of the depression into a pond and to shape the rest of the depression, to where it runs under the fence, so it would form a seasonal creek-bed. This arrangement captures water in the pond as it flows in from the eastern boundary. Once the pond is full, the water continues flowing along the creek-bed and under the fence.

When it was time to plant, holes were cut in the weed mat (see below), a little coarse sand was worked into each hole for drainage purposes and a handful of compost for nutrients.

Development of creek-bed area: laying non-woven weed mat in the creek-bed (a); spreading coarse sand over weed mat (b, c); stones are laid on top of the sand in the creek-bed and weed mat has been laid around existing young plants (d, e); free mulch from the tip has been spread over the weed mat around plants along eastern fence (f); pond is full, stony creek-bed formed and mulch spread (g); the young plants are developing nicely (h). Compare this with the photo on the previous page.

Plants

We need to keep an unobstructed water flow here when it rains heavily, so there are few plants in the actual creek-bed. A few sedges, common Billy buttons and a couple of dianellas snake through the centre. Closer to the edges of the pebbled creek-bed are some sweet bursarias, prostrate melaleucas and callistemons. The water seldom rises high enough to flood these latter plants and so far they have shown no ill effects from their occasional dunking.

Mulch used

The whole area from the creek-bed to the eastern fence was covered with a commercial non-woven weed mat, said to last up to eight years if not exposed to sunlight. This mat needs to be weighed down with timber or bricks as soon as it is laid or it will blow away before you get the top mulch on. Coarse river sand was spread over the weed mat in the creek-bed for the pebbles to nestle into and to prevent them from wearing holes in the weed mat. Over this went barrow load after barrow load of the same river pebbles used for the succulent plot.

Watering

This area is watered with a hand-held hose and occasionally, a sprinkler. By the end of the second summer watering was done once a fortnight. By the fourth summer I was forced to leave this area unwatered and it continues to thrive.

Evaluation

Very few weeds have taken root in the sand and are easily pulled out. So far none have

The local duck population, many other birds, insects and frogs, all enjoy the pond and surrounds.

grown through the weed mat. At first the pale pebble mixture that was available from a nearby quarry looked very bright and stark, but now that the plants have spread and there is a litter of fallen gum leaves scattered over the stones it looks more natural.

Ducks, frogs and many water-based insects make use of the pond when it has water in it and wattle birds and magpies bathe there. We were not interested in having a pristine crystal-clear pond that would need constant additions of a variety of products to keep it that way, so we haven't lined it, being reluctant to lose the soil-water interactions that naturally keep the water in a healthy condition and full of life. However, the pond quickly loses water, especially after summer rain. We added bentonite (sodium bentonite, available from rural supply stores and sold as an animal supplement) to the water to seep into the cracks in the soil and it has slowed down the leakage. We did know it would be a seasonal pond, but were hoping for a slightly longer season. Even so, the whole area looks great.

Between creek-bed and eastern fence

This area of native and indigenous shrubs and grasses is a natural extension of the creek-bed. The plants will eventually hide that section of fence and create a habitat area.

Preparation

The weed mat used in the creek-bed extends to cover this plot as well. I already had some shrubs planted by the time I obtained the weed mat, so it was a bit of a fiddle to lay it over the small plants and cut around it. Prior to planting, the area was mown and individual holes were dug, with a handful of compost being worked into each hole. It was a simple matter to cut holes in the weed mat to complete the planting of this area.

Mulch used

The weed mat described above was covered with partially decomposed mulch collected free of charge from the tip – a shredded mixture of whatever green waste is dumped there. Eventually, the area will be self-mulching from leaves and flowers shed naturally and from fine prunings I will let fall when I trim plants to keep them bushy.

Plants

We have a lovely view to the east, so didn't want to plant anything on this side that would end up growing to block it. All plants chosen for this plot will grow to less than three metres. They include both prostrate and shrub wattles, callistemons, grasses, a grass tree, a kangaroo paw and a Western Australian feather plant. The planting layout has resulted in a shrubbery that allows space to wander through, with prostrate species spreading around the base of taller shrubs to act as living mulch. The existing mature red gum casts dappled shade over the area for some of the day, and drops its bark and leaves to replenish the mulch.

Watering

Watering was initially with either a hand-held hose or a sprinkler. No watering system was installed here, as it was my intention to gradually wean the plants until they eventually had no extra water. This happened, of necessity, in the fourth summer.

Evaluation

A few weeds sprouted, apparently from the tip mulch, but were easily pulled out. This plot is coming along well and was soon attracting small birds that peck around the shrubs and lizards that dart through the mulch.

Front and side of deck

The front-of-deck plot began with a very clear notion that we wanted palm trees to frame the view and shade the deck. We erected lattice on the north side of the deck and built up a bed there with the aim of growing two four-grafted fruit trees trained onto the lattice, espalier-style.

Preparation

The strip in front of the deck was dug lightly to aerate the soil and remove weeds. Topsoil and a scattering of compost were raked over the top. So there wouldn't be an ugly view of under-deck stumps, we cut some

Development of creek-bed/eastern fence area:
pots are set out in place prior to planting;
different views showing how the plants in this area
have developed.

Espaliered fruit trees on side of deck.

Front-of-deck bed with leftover roofing iron as a backing. I love the contrast of the blue and red.

leftover Colorbond roofing iron and screwed it to the front of the stumps as a backing for both beds. The north-facing bed had more generous amounts of compost and some decomposed chicken shed litter worked into it in preparation for the fruit trees I'd ordered.

The beds were edged with leftover bricks. Because the house bricks are the type with holes through them, not the most appealing look, we eventually put matching colour paving tiles along the top over the holes, and a small bright blue ceramic tile, diamond-wise, in the centre of every second paver. It looks great! What's more, it cost practically nothing because the bricks and the ceramic tiles were leftovers and the pavers were dirt-cheap remnants from a local clay works.

Mulch used

Initially, the plots were mulched with sugar cane mulch and a scattering of compost. This

It wasn't long before the trees were producing luscious fruit.

has been replenished with pea straw and prunings from the annuals planted around the base of the palms and herbs around the fruit trees.

Plants

The starting point for the front-of-deck plot was the palms. After researching the subject and consulting our preference for smooth-trunked palms, we decided on bangalow palms. These did not survive the unusually severe frosts of the first winter, so had to be replaced. This time we looked around town to see what palms were thriving in the climate and decided on cocos palms – these can be a weed in places further north, so avoid them in your area if this is the case.

To fill in around the palms I chose a blue/mauve-flowered pansy, normally an annual, but it self-seeded and lasted about three years after a couple of heartless prunings when they became straggly. The contrast of the blue/mauve flowers against the red Colorbond background was quite dramatic.

The plot needed some plants that would be between the pansies and the palms in height, so Siberian irises (purple/blue-flowered) were added.

I had brought with me from the old garden some pots of a very tasty but tiny-fruited alpine strawberry which made an attractive, edible edging plant at the front of the bed.

The grafted fruit trees, two varieties each of plum, nectarine, peach and apricot, were ordered from a nursery in NSW that specialises in growing multi-grafted fruit trees on dwarfing root stock. My plan was that

by growing small trees on lattice close to the house we would avoid having large gluts of fruit, much of which would be wasted. The proximity to the house should discourage avian predators. If not, it will be easy enough to throw a net over the lattice to protect the crop. In front of the two fruit trees I've planted calendulas, borage and nasturtiums. These will self-seed continuously, give me colourful flowers that will attract beneficial insects, and provide free mulch.

Watering

Watering of both beds is by microsprays every few days, depending on weather conditions. During the 2006–2007 drought the watering changed, first to dripper hose and then to hand watering with grey water.

Evaluation

As mentioned above, the first lot of palms had to be replaced, even though my research had indicated they should be suitable. The locals claimed it was a more severe winter than usual, with more frequent and colder frosts, so perhaps replacements of the same species would have survived. I wasn't willing to risk losing my money twice and being delayed in the development of the plot, so opted for a different species. Apart from that setback, we are very pleased with the look of this plot and its relaxed tropical holiday feel.

The fruit trees are growing beyond my expectations, some fruiting in the first year. There was some snail damage that I'll have to be vigilant about checking and dealing with in future.

Edging being installed for vegie beds.

Vegie garden

Once the huge pile of pebbles that was dumped here had finally all been moved to its allotted locations, work could begin on the vegie garden in its protected microclimate at the north end of the house.

Preparation

Weeds and grass were brush-cut in two strips about 90 cm wide from the corner of the western and northern fence lines and a hexagon shape in the approximate centre of the area. Newspaper was laid over the ground and red gum sleeper edging laid so the paper

Topsoil in the centre bed.

Centre bed planted, mulched and irrigated.

was tucked underneath it. Beds were built up with bought topsoil. Decomposed manure, compost and Dynamic Lifter were spread over the topsoil and raked in.

Peas were planted as a green manure-cum-mulch in the western strip.

A weed-free walkway was made of the grassy, weedy area surrounding the vegie plots by covering it in heavy duty builders' plastic with more of the contractor's shredded tree mixture on top.

Mulch used

When the peas reached flowering stage they were pulled up and left on the surface as mulch. Lucerne hay was spread over the pea straw and on the remainder of the beds.

Plants

Our favourite seasonal vegies grow in most of these plots. For convenience, and to assist future crop rotations, I decided to regard each fence bay as a separate bed. The bed in the corner where the fences meet forms a triangular shape and has been devoted to a

Constructing a weed-free walkway: I wanted to eliminate the mowing job around the vegie garden (a); black builders' plastic was spread over the cut grass (b, c); a load of shredded tree mulch was spread over the plastic (d); job complete (e); overview of the area with mulched walkway complete (f).

The first healthy young crop of corn and peas.

Herbs in the corner bed with lettuce at the front.

selection of permanent herbs: rosemary, thyme and lemon grass. The last bed on the northern fence has a wire trellis (leftover reinforcing mesh) on the fence and is another bed of permanent plants: youngberries on the trellis, asparagus in front of them and marjoram and French tarragon at the front of the bed.

The hexagon bed has a four-grafted citrus tree in the middle, grown on dwarfing rootstock so it will reach about two metres high and wide. I began by planting strawberries to define six separate triangular

beds radiating out from the citrus tree, but the strawberries liked the idea so much they decided to spread into the adjacent beds. I lost my pretty design, but gained enough strawberries to feed several families.

Watering

Both strip and hexagon plots are watered by a microspray system, cleverly designed so that each bed has a separate tap, allowing precise control of watering. In practice, it is usually the case that the strip beds are all watered together and the hexagon is watered all at

Broccoli at the back with red cabbage in front – aesthetics and edibles all in one go!

Productive young tomato plants.

Zucchinis and lettuce interplanted.

once. The separate taps are most useful when new seedlings or seeds have been planted and need watering more frequently than adjacent more advanced vegies. During the drought I laid some dripper tube, which proved effective until Stage 4 restrictions were applied and I could no longer water, so let the vegies go for the remainder of the summer.

Evaluation

From the first season the plots have been very productive, despite some unexpected pest problems. Snails and vegetable bugs were very prolific and it took me some time to decide on and implement control programs. Now I recognise vegetable bugs for the pests that they are, I squash them mercilessly at every opportunity and spray pyrethrum on clusters of baby bugs that are hard to reach. Snails are squashed, deterred by shell grit sprinkled around seedlings, drowned in beer traps or baited with one of the snail baits that doesn't harm pets or wildlife. In addition, I've encouraged beneficial predators into the vegie garden area by planting some flowering herbs amongst the vegies and allowing many plants to flower and seed, and by installing a small shallow pond behind the rosemary bush in the corner bed.

Shed surrounds

The east and south-facing walls of the shed gave me some shade and wind shelter, which I thought would be a suitable spot for a few less hardy species I'd brought from the old garden. In addition, I thought the plants would nicely soften the shed.

Preparation

The excavation of the site for the shed had left the surrounding area weed-free so there was no need to dig. The sand for the creek-bed plot had been dumped nearby and there was a little left when the plot was finished, so I raked it over into narrow strips beside the shed. A few barrow loads of topsoil and a scattering of compost were tipped on top and raked in. The bed was edged with leftover bricks.

First planting at rear of shed consisted of plants from the old garden.

When a path was eventually installed from the house to the shed it left a nice triangular section bordered by a section of shed wall, the path and a lattice fence separating the vegie garden from the main garden. This was ideal for another small plot. Weeds were dug out and, as it was very heavy clay, much compost, mushroom compost and soil improver and mulch were dug in along with some topsoil.

Mulch used

Sugar cane mulch was used initially and has since been augmented by prunings from the plants in the bed, finely shredded eucalyptus mulch and lucerne hay.

Plants

Tree dahlias, windflowers, columbines, hollyhocks and a hydrangea were planted in the strip beds at first. The triangular bed is home to three Standard 'Burgundy Iceberg' roses, with *Nepeta* 'Walker's Blue' and peppermint spreading around their bases as mulch and attracting a host of beneficial insects.

The 'Burgundy Icebergs' are flowering and the nepeta and peppermint are spreading and attracting beneficial insects.

Watering

A microspray system waters both strips and triangular plots; 15 to 20 minutes every second or third day is usually enough. Hand watering with grey water became necessary during the extended drought.

Evaluation

In the first summer the plants on the eastern side of the shed appeared to be more stressed than I had expected during hot days. I put this down to reflected heat from the shed wall. As long as they got enough water, these plants thrived, but only in fits and starts during the milder seasons. Severe frosts froze them and intense heat in summer scorched the new leaves. They also needed more frequent watering than most others in the garden. I hardened my heart and replaced these sensitive specimens with some tougher species: cushion bushes and red cordylines.

I am particularly pleased with the rose/nepeta/peppermint combination; the

The rose corner when first planted.

The new, hardier plants at the back of the shed are cordylines and cushion bushes.

'Burgundy Iceberg' roses and the soft drift of lavender-coloured nepeta and peppermint flowers complement each other beautifully, and the swarms of insects around them are a delight to a biodiversity-conscious gardener. The fresh peppermint tea every morning is an energising bonus.

Between house and creek-bed

A small bed in this plot took shape very early in the garden's development, but it was some time before a clear picture developed for the rest of the area. Initially, I'd had vague concepts of a native walk-through shrubbery, but it had no satisfying concrete shape until the path was installed along the eastern side of the house, bordering the front-of-deck plot and leading to the shed. Then a quite different concept emerged, still with mainly native shrubs, but with a definite structure and design.

Preparation

The whole of this area was quite heavy clay with poor drainage and little organic matter.

To begin improving the texture of the clay, and in an attempt to keep down the dust in that very dry summer, almost the first thing we did in the garden soon after moving into the house was to spread a truckload of mulch (sawdust-based stable bedding) over this whole area. The mulch was free from a nearby horse stud, but we had to pay for transportation.

The first bed to be planted in this area was built up in a roughly oval shape on the site of the sand pile. There were few weeds and these were weak-rooted and easy to pull out after having been buried under the sand for several months. I raked leftover sand and stable mulch to form the shape I wanted then added topsoil and decomposed manure and raked it into the sawdust/sand mixture. The bed was edged with leftover bricks.

Development of the remainder of the area did not take place until over a year later, after the path was installed. By this time the sawdust mulch was decomposing nicely, but there

The first area to be developed between the house and the creek-bed was planted with silver birches, native shrubs, irises and self-seeded poppies.

were many weeds established. I spread gypsum over the area to break up the clay. The weeds were mowed closely and a roughly circular outer bed of topsoil was built up to enclose an inner flat area. The outer circle meets the path at two points so that the path forms one section of the rough circle. The inside edge of the raised bed was shaped to form flowing curves, making the width of the bed variable and defining a flowing curved shape inside. This flat area inside the raised bed was to become a herb lawn with a double-grafted citrus tree in the centre. The design was planned so that it would eventually form a garden room with the plants in the outside raised bed becoming the walls and the herb lawn the fragrant floor.

The planting hole for the citrus tree in the centre of the herb lawn was prepared months before the tree was planted. I dug a large hole, breaking up the underlying claypan with a crowbar, and worked in enough topsoil, compost and decomposed manure to raise a small circular mound that the citrus tree would later be planted into.

I had to break up the claypan beneath each planting hole in the raised bed with a crowbar. The planting holes were enriched with compost and decomposed manure.

The outside of the raised bed was edged with a variety of leftover materials including bricks, tiles, lattice, roofing iron and rocks. Several square and rectangular blue-glazed pots of succulents are inserted as part of the edging. A friend calls it my 'whimsy'. It cost practically nothing, used materials that would otherwise have been wasted and looks unique and interesting.

Mulch used

The first oval-shaped bed was mulched with sugar cane mulch. The raised circular bed had the last of the truckload of shredded tree mixture spread around after planting and the flat area for the herb lawn was sheet mulched with newspaper covered with pea straw. Planting holes were cut through the sheet mulch and a handful of compost added to each.

Plants

In the oval bed two silver birches give height, and will eventually provide mulch and shade for the plants beneath. Golden diosmas and a ground cover thryptomene are spreading beneath the birches to keep the root systems cool and mulched and to provide interest lower down. A border of irises defines the edge of the bed. Around the diosmas and thryptomene changing seasonal interest is provided by daffodils, Siberian irises and scattered, self-seeding poppies and hollyhocks.

The raised circular outer bed is planted to a wide variety of hardy, mainly native shrubs up to two metres tall, with ground covers spreading around their bases. The inner curving edge of the bed is defined by irises and tritonias. Near to where the bed meets the path on both sides I've planted crepe myrtles, which will grow to about four metres and eventually give some shade to the deck. The plan is that they won't grow tall enough to block the view, will let light through in the winter, flower continuously for months in the summer, drop leaf mulch on the bed beneath and have a graceful branching structure to provide visual interest during the winter.

Preparation and development of front of house area: looking over circular bed and herb lawn from deck soon after area was planted (a, b) (note frost protection around citrus tree in photo b!); section of the circular bed from a different angle (c); same angle as above after plants have grown a little and started flowering (d); different section of circular bed showing crepe myrtle in flower (e); Mary on the deck enjoying the garden (f); different angle of circular bed (g); irises in flower and herb lawn spreading (h, i, j); overview of circular bed and herb lawn from the deck (k).

The herb lawn is planted with pennyroyal, lawn chamomile, a ground-hugging thyme called *Thymus* 'Doone Valley' and pennywort. A creeping native violet from the adjacent raised bed soon began to spread among the herbs.

Watering

The oval bed is watered by microsprays. The raised circular bed has a dripper system installed, with a separate, adjustable-flow dripper for each plant. The herb lawn has a length of soaker hose snaking through it. The citrus tree is surrounded by lawn chamomile, which doubles as living mulch. This inner circle is sometimes watered with a sprinkler. In addition, I've sunk four lengths of wide-diameter pipe around the base of the tree (see Chapter 7) and once or twice a week these are filled with water that seeps down to the root zone.

Evaluation

The only problem has been some weeds growing through the newspaper sheet mulch. This was solved by digging them out and re-covering the spots with thicker wads of paper or cardboard, a simple if time-consuming procedure. If I ever do this again, I'll wait until I have access to thicker cardboard, or buy some weed mat. Most plants have flowered and are growing well, attracting small birds and beneficial insects. I love the look of the leftover edging, but have found that its uneven nature provides lots of hidey-holes for snails. Now that I know where they are though, those same hidey-holes have become snail traps.

The herb lawn was cut to the ground and covered with sugar cane mulch in the summer of 2006–07 because there was no water to keep it going. If it fails to regenerate from the roots, I'll replace it.

Northern fence

When I thought I'd planted all the plots I intended to plant, I discovered I still had a variety of plants left, most of them given to me as bulbs or cuttings. The only space available was a section along the northern fence line.

Preparation

I decided to make this a no-dig bed. A curved strip was brush-cut and blood and

No-dig bed along the back fence with red sages in flower.

bone scattered over it. Thick layers of newspaper were spread and covered by alternating layers of pea straw, bagged topsoil, Dynamic Lifter and compost, finishing with a layer of pea straw. Planting holes were made in the pea straw, mixing in the other materials, and a handful of compost and topsoil was added to each.

Mulch used

The built-up mulch, as above, forms the bed.

Plants

There was initially a lovely hardy mixture of sages, osteospermum daisies, verbena, vallotta lilies, columbines and erigeron.

Watering

Watering is by a length of soaker hose through the middle of the bed.

Evaluation

No problems for the first few years. For an add-on plot it looked as if it was always meant to be there. However, I decided that this area was one I would let go during the drought and have since replaced many of the plants with hardier wattles and grevilleas.

Extras

Rocks

The large rocks you see in some of the photos were obtained as part of a barter deal with a contractor who had the required equipment and access to a rocky site on a property where a dam was about to be built. He was kind enough to collect, deliver and place the rocks,

Rocks delivered and positioned by a contractor. Don't try this yourself unless very brave and strong!

which we could never have managed on our own. Most of them are around the pond. A cluster of three that were placed as a feature on the east of the house enabled me to plant a palm in the centre and kangaroo paws and ground-covering common everlasting native daisies around the outside.

A number of smaller rocks from the same source were used as part of the whimsy edging around the circular bed on the east of the house.

Lattice

As well as being used to extend the height of the western fence and as a support for the espaliered fruit trees, lattice was used to make a short length of fence between the house and the shed. This separates the vegie garden from the main garden and, as we have no front fence, gives us a secure yard to leave the dog in when we are away from home.

Paving

The path along the east of the house is concrete pattern paving, terracotta red in colour to blend in with the house bricks, with a simple brick-like pattern. It reaches from one end of the house to the other and continues to the shed and the gate in the lattice fence. Variations in width and some curves in the layout add interest to this very functional feature. This was the most expensive part of the whole garden project.

Shed

As we needed storage space for tools and building materials, the shed was built before the house, of Colorbond metal that would match the house colours. With the help of our architect we took a 'guesstimate' of the best place for it in relation to the house and future garden and it has worked out quite well, especially now there are a few plants around it to soften the hard lines.

Recycled fence art

Most of the fences will eventually be hidden by plants. However, a two-bay width had to be left with no plants in front of it to allow future access to a septic inspection pit, and to prevent roots growing into the area. After a great deal of discussion about what to do with this spot, we decided to create some fence art. The two fence bays were painted bright yellow. Leftover red roofing iron was cut to represent hills and attached to the bottom of the fence. The cut-out sections were reversed to show the silver colour and inserted to look like more distant hills behind the red ones. Finally, we bought a unique 'bird' made from recycled tractor parts from a local sculptor and stood it in front of the fence. We're extremely pleased with our fence art; it cost very little, makes use of leftover materials and is certainly a talking point.

New lessons

No matter how much experience and knowledge you bring to any project there are new lessons to be learnt along the way. I began this garden confident that I knew what I was doing, and most things worked out as planned, but I did learn some worthwhile lessons.

- Be careful about what you plant close to metal fences or other structures; the

Paving: concreters are preparing to lay the path (a, b); concrete being poured, note reinforcing mesh in the bottom (c, d); pattern template being applied to the wet concrete (e); red oxide being scattered over the path (f, g); job nearly done (h, i).

reflected heat can damage young and tender plants. I suggest you choose tough species and more advanced plants over very small tubestock for these locations, and make sure they are hardened off before planting. It might help too, to plant in autumn so the new plants have maximum time in the ground to toughen up before the next summer's heat.

- Town gardens have just as many pests to deal with as country gardens, only they are more likely to be snails and vegetable bugs than rabbits, 'roos and grasshopper plagues. Be observant and deal with pests as soon as you see them – don't give them the opportunity to build up. At the same time be aiming to increase the biodiversity of your garden, as described in Chapter 2, so that before long you'll have a range of birds and beneficial insects to keep pests in check.

- Commercial weed mat and thick cardboard from white goods boxes are much better sheet mulch than either newspaper or thin cardboard boxes. Be patient and wait until you can buy weed

We are pleased with our fence art using recycled materials.

mat or collect the appropriate boxes. Newspaper does work, if you use enough of it and are vigilant about digging out weeds that break through, and re-covering the spots.

For the future

Any garden is always a work in progress. We have now reached a stage we are calling finished, but are aware that there is still room for change and improvement. A front fence is a possibility, and that will bring new ideas and planting opportunities around it. We plan to build up a new bed in a section of the lawn and plant some bushfoods. We have many ideas, some of them perhaps too fanciful, for enhancing the pond area. There is an ideal spot near the shed for a small water feature, or perhaps it would be more practical to install a rainwater tank there (the tank won). Inevitably, plants will die or outgrow their space and need to be replaced or moved to another location. In addition, there is always innovative garden art to search for to add intrigue and individuality. The hard yakka is done, now there's time for relaxation and fun. I wish you a satisfying journey as you create your garden and many years of enjoyment with the end result.

The bulk of our garden was established during 2003–04, after a severe drought. Little did we know that just a couple of years later there would be an even more severe drought, which would test my strategies and plant choices and teach us some new lessons.

DROUGHT PROOFING

We moved into our house in December 2002, during what was widely considered to be a record-breaking drought. Four years later, in 2006–07 there was an even worse drought. Residents of our town were subject to Stage 4 water restrictions for months, including the hottest, driest months, and had only grey water from the house to bucket onto the garden. Our town even had the dubious distinction of completely running out of water for a brief period. How did my very young garden survive? Surprisingly well.

To begin with, I had intentionally designed the garden to be drought-tolerant; however, I had not expected it would need to be quite so tolerant at quite such an early stage. The

The fruit trees thrived on their grey water rations.

2006–07 summer was planned to be the last summer of regular watering for some areas, after which they would receive only occasional deep watering during extreme dry spells. The problem was that we had an extreme dry spell and no water was available. Many plants had to make it on their own, or not. Most came through very well, simply because that's the way I had planned it.

Climate change and your garden

The overwhelming opinion of the reputable scientific community is that although there have always been variations and natural cycles of change in climate, human-induced climate

At the end of the drought the herb lawn had died, but nearly every plant in the circular bed survived.

change is happening. We are living with it right now and will experience more frequent severe droughts of the kind we saw in 2006–07 into the future. One of the consequences, and an important one for many in our garden-loving nation, is that we are all going to have to be smarter in the way we manage our gardens. Predictions are that there probably will be less water available in most populated areas of Australia, so we need to implement strategies that will ensure our gardens of the future use less water, but still thrive. A big ask, you might think. But I believe it is not too difficult to achieve. Perhaps the most difficult aspect is going to be to change the still very prevalent mindset that desires the beauty of sweeping green lawns, European trees and exotic floral displays in our home gardens.

There are two aspects to drought-proofing your garden. The first is long-term planning and preparation that are built into your garden from the beginning and maintained permanently. The second is putting in place short-term measures to get through the current drought, whenever it might be. Of course, these two aspects will overlap.

Planning and preparation

If you have followed the advice in this book thus far, your garden will already be in a good position to survive tough conditions.

The soil reservoir

Your soil acts as a huge reservoir, holding more water than you would imagine. But two conditions must prevail in order for this to happen. The soil must have a good open structure that allows water easily to percolate through, and there must be plenty of organic matter in the soil, which acts as a sponge, soaking up and holding the water where it is available for plant roots.

Sand, clay and organic matter

Sandy soil has too open a structure with little or no organic matter. Water passes through it very quickly and there is no organic matter to absorb and hold it. The fine particles of clay soil are often bound so tightly together that the soil is compacted, especially if it has been subject to foot or machinery traffic, and there is no space between the particles and aggregates for water to access (see Chapter 1 for more details). Heavy rain often causes it to become waterlogged and when it dries out it bakes hard like – well, clay.

In the preparation stage of your garden – and when replanting where there have been casualties during a drought – before planting anything, get as much organic matter into the soil as you can. Materials to use include any decomposed manures, compost and decayed or decaying plant matter. Very sandy soil can have clay added. Dig it in. You know by now that I don't like, or usually advocate, lots of digging, but this is important to the future survival of your plants and you only need to do it once. Very clayey soil can have sand dug in to open it up and allow water penetration.

I heard a television garden presenter some years ago saying that if you add sand to clay soil all you get is clay with sand in it. To some extent this is true, but it oversimplifies what actually happens. What you get is a mixture of particles of different sizes – the very small clay particles and the larger sand particles – and the particles cling together into different sized aggregates. This means there is more space between the particles and aggregates for air and water to permeate. Add organic matter to soak up and hold the water and you are well on the way to having good gardening soil.

The addition of organic matter to all soils and adding clay to sand and sand to clay will give you soil aggregates of differing sizes and shapes, thus allowing plenty of spaces through which water can permeate. In addition, the organic matter will increase the soil's capacity to hold water. There are any number of soil wetting agents available, which are claimed to help hold moisture in the soil. I seldom use them, and when I do it is mainly in pot plants. They are expensive to use throughout the garden and need repeat applications. I much prefer to use organic matter, which is a natural part of the ecosystem, often freely available, improves the soil structure and biodiversity and provides nutrients as well.

Raised no-dig beds

If you really don't want to dig, and I seldom do, build up raised no-dig beds as described in Chapter 5. There are numerous benefits to no-dig beds, but the most significant with regard to drought proofing your garden is their wonderful water-holding capacity. When building a no-dig bed over compacted clay soil, it will be necessary to incorporate a bottom layer of coarse material such as gravel to ensure good drainage. If the clay becomes very waterlogged during the wet times, install some drainage around and/or under the bed to allow water to flow away from it.

Here I am building up a new raised bed for a bushfood garden. Plenty of organic matter was raked into the imported topsoil.

Over time, the activities of all the soil biota that will thrive in the raised bed will incorporate the decomposing materials in the bed into the underlying soil, thereby improving it.

Aeration

In some cases – lawns or slightly compacted existing beds, for example – it can be useful to aerate the soil physically. Walking around a lawn with spikes strapped onto your feet is an easy way to create holes for water to enter. A garden fork can be used to aerate any bed when necessary.

Mulch

No garden bed is complete without a surface layer of mulch, the many benefits of which were discussed in Chapter 6. Keeping beds, and individual plants mulched, and renewing the mulch frequently, is a valuable drought-proofing strategy.

- Mulch slows down evaporation of moisture from the soil, thus keeping moisture in the soil reservoir for longer.
- Mulch protects the soil surface from the harsh sun, keeping the soil (and plants' roots) cooler.
- Mulch prevents soil erosion when sudden summer rain storms or strong winds occur.
- Mulch slowly decomposes and is incorporated into the soil, providing nutrients and increasing organic matter.
- Mulch encourages the activity of numerous organisms in the soil, which burrow down and open up channels through which air and water can enter.

Mulch in dry times

I have read and heard criticisms of the use of mulch during dry times on the grounds that it stops light rainfall from reaching the soil. This is indeed the case, but it is not a reason to abandon mulch. Rainfall light enough that it will not penetrate beneath the mulch only dampens the surface of the soil anyway and doesn't reach the plants' root zone. In addition, it will quickly evaporate in the sun and wind. A light rain that dampens the mulch, on the other hand, will readily be absorbed by it and create a beneficial humid and cooling zone around the plants for some time.

When all the above benefits are taken into account I consider that they far outweigh the supposed disadvantage. How to get water beneath the mulch will be discussed in the section on short-term drought-survival strategies.

Select drought-surviving species

Your choice of plants is the next critical factor in drought proofing. Features to look for in drought-tolerant plants were discussed in detail in Chapter 7. Learning how to choose drought-tolerant plants suitable for your garden is one area where I cannot overemphasise the importance of trial and error, observation and information sharing.

Drought-tolerant once established

Indigenous species are likely to survive and thrive in your garden with little or no extra water, once established. And this is the point to bear in mind with all plants that are recommended as drought-tolerant – it only applies once the plants are established. Most of the plants you will have in your new garden are the babies of the plant world and need appropriate care, and water, if they are to survive to reach their hardiness potential. In a natural ecosystem many more seeds are dropped than ever germinate and many more seedlings begin to grow than ever survive to maturity. Pests, diseases and adverse weather conditions all take their toll. This is not what we want to happen in our gardens. Don't think that because a popular and authoritative garden presenter recommends a plant as being drought-tolerant you can plant it and forget it. This strategy is far from a guarantee of success; all new plants need care to ensure their survival in the garden.

Survival in the plant world

When we talk about plants surviving drought, we need to know what survival means in the plant world. All plants have evolved strategies to ensure their survival as a species; some of them are more appropriate for our gardens than others. Some trees under stress will drop numerous leaves so the plant needs to provide less water for those remaining. This is not a bad strategy for a garden plant; the plant lives and it drops some free mulch that benefits other plants nearby as well. Some plants drop limbs when under stress. This is not such a handy device and could cause problems in the garden, depending on what is under the limb when it comes down. Some plants produce copious amounts of flowers and seeds and then die, dropping the seeds, which will germinate at some indeterminate time in the future. This might ensure the survival of the species, but it will not contribute to the aesthetic appeal of our gardens. Others might

appear dead above the ground, but will shoot again from an underground root system when it rains. How do we learn about these survival strategies so we can choose plants appropriate for the garden? Trial and error, observation and information sharing.

Timing

Timing is important. Historically, the season or two after a drought usually have average or above average rainfall. How much we can rely on historical data into the future is unknown, but until trends are clearer it at least gives us a starting point for our decisions. This is the time to plant new areas or to restore drought-ravaged gardens to ensure the maximum time elapses until the next drought. By then your new plants should be established enough to get by with an occasional drink of grey water. By the second drought, and probably well before, your hardier plants should be able to survive without any extra watering.

Of course, this all depends on your situation. If you are in a very dry area, subject to frequent severe water restrictions, you will want to find plants that can survive on neglect from an early stage. This can be done by trial and error, observing your own garden and your neighbours', and by sharing your observations. When you find species or varieties that thrive in spite of every setback the weather throws at them, plant more of them and replace more fragile varieties with them.

Learning and sharing

In my case, I often share observations with friends and neighbours, but I always let them know that my soil is very clayey, so species I have success with might not suit them if their soil is sandy. A friend who lives two blocks away has very sandy soil, for example, so many of the plants that thrive in my soil, which is clay-based but now enriched with organic matter, will not suit her different soil conditions, even though they are exposed to the same general climatic conditions. The same applies to lists of recommended plants published by garden experts. Consider them to be a useful guide, but not a substitute for trial and error, observation and sharing information within your area.

Even within one garden, plant survival can be variable. I had three *Eucalyptus caesia*, 'Silver Princess' trees. One thrived, one struggled through and one died. Of four *Erica darleyensis* shrubs, three died and one is still thriving. The plants I had a great deal of success with were wattles, grevilleas and callistemons. Every variety in my garden thrived through severe drought, most with no extra water. Indeed, one indigenous wattle – *Acacia acinacea*, gold dust wattle – not only kept growing, but I found tiny seedlings self-propagated under the two original plants. These were transplanted to replace the less reliable ericas.

Despite the fact that every young callistemon in my garden survived well without any watering, I noticed a number of quite large callistemons in front gardens and on nature strips that had died. Another important aspect of species selection is to try to find out what a plant's natural growing conditions are. The natural habitat of many varieties of callistemons, for example, is a shady gully or close to a river or creek bank. Though they are tough plants that will grow in a range of habitats, when conditions vary too far from those found in their native habitat, they will die. An exposed nature strip is not an ideal spot, long term, for a shady gully plant.

Sometimes information about a plant's growing requirements can be found on the label. If you have knowledgeable local nursery staff, they are another source of information. Experienced gardeners near you are usually willing to share information. In addition, I recommend you buy yourself a good garden encyclopedia, or ask for one as a gift.

See my plant list in the appendix for more details about plants that survived in my garden. Some plants that did well in this area in general include lavender, rosemary, roses, pelargoniums, diosmas and irises. Bulbs and corms will usually reappear at the appropriate time. Seeds that have been dropped from the previous season's flowers generally remain in the soil until conditions are right for them to germinate. This does include weed seeds as well, so be prepared to get straight onto weed control as soon as the drought breaks.

Succulents have a well-earned reputation as being drought-tolerant. However, as for other types of plants, they are only drought-tolerant once established. Do not expect newly planted succulents to survive a drought without water. Be aware, too, that most succulents are frost tender, so that will limit where you can successfully grow them.

Design for drought proofing

Conditions vary from plot to plot within a garden. You probably have a shady wind-protected area, an exposed spot in the path of strong winds, a hot dry area, and beds that are protected by fences and buildings. All these variables are known as microclimates within the garden. Make sure you match up a plant's needs with the area you place it in. Observe the existing microclimates in your garden and create new ones to suit your own needs. If you really want to grow plants that need dappled or dense shade, you must provide the shade, whether it's from existing buildings, specially erected structures or from hardier large plants.

Wind protection

Wind is very damaging to a garden and the strong hot winds frequently experienced during droughts even more so as they increase evaporation, whipping the moisture from both plants and soil in their tumultuous progress. If you have plots exposed to strong winds, you need to do something to reduce the wind's impact. This might be a fence, though not a solid fence, which will only increase the turbulence of the wind on the other side; lattice; a shadehouse; a windbreak of hardy plants to protect the more fragile specimens, or a temporary shadecloth barrier.

Group similar plants

Group plants with similar growing requirements together in the same plot and place those that need a higher degree of care in plots closer to the house. This makes all maintenance tasks easier, especially the drought job of bucketing water to them to keep them alive. As well, the house itself creates shade at different times of the day and provides some wind protection, depending on its orientation and the direction of prevailing strong winds. The hardiest, most drought-

Drippers are very water efficient, delivering water right where it is needed with no wastage.

tolerant species can be furthest from the house. Try to position them so their shade is cast on areas that need it and they can act as windbreaks.

Watering systems

A variety of watering strategies, including using grey water, were discussed in detail in Chapter 7. The most water-efficient methods are dripper systems, of which there are a few different types available, and porous or leaky hoses made from recycled rubber. These will serve you well until your area is put onto Stage 3 and 4 restrictions. Stage 3 restrictions allow you to use dripper systems, but at such limited hours that it is often impractical for large gardens. Stage 4 restrictions forbid the use of all garden watering systems. This is when the challenge really starts and it is time to move on to short-term strategies to carry your garden through the current drought.

Tanks

The final suggestion I have as a long-term strategy is to install rainwater tanks, as many and as large as space will allow and you can afford. Then, when restrictions are introduced you are independent. There might not be much rain to top them up, but even in droughts, thunderstorms do happen and if there is any moisture in the air, dew will form on your roof catchment and trickle into the tank each day.

Sailing through a drought

With your garden set up as described above, you are in as good a position as possible to sail through drought conditions with minimal losses. If you do lose plants, consider it a learning experience and replace them with varieties that did survive.

The vital thing to do now is get as much water as you can to those plants that need it the most. It is important to ensure that the maximum amount of water possible reaches the root zone where it can be useful to plants. This means that when using dripper hoses you place them under the mulch so they are in contact with the soil. The mulch will help keep the water in the soil by slowing down evaporation.

Stage three

Under Stage 3 restrictions in Victoria (other states might differ) you can use dripper systems and hand-water on two days a week between set hours. With a small garden this might be sufficient. If not, implement the strategies discussed below for Stage 4.

Stage four

Once Stage 4 restrictions are in place, we need to be a little inventive and very persistent. Carrying buckets of water from the house is virtually compulsory, unless you have installed a reticulated permanent grey water system throughout the garden, and not many of us have. Some new housing estates have built this feature into the development, along with water-efficient appliances in the houses, and I'm sure these factors will play an increasingly important role in home buyers' decision making in the future.

Make a plan

As soon as you are notified of when Stage 4 restrictions will apply, make your plan. Decide which areas or individual plants you must water, and which you think will survive on their own, or are willing to sacrifice. These decisions will not necessarily be made on purely practical grounds. There might be a plant that a special person gave you or one that brings back memories or has particular meaning for you, or that you just love so much you are willing to put the effort into pampering it to ensure its survival. Make a list of plants you intend to water.

Next, decide where the water is going to come from. Along with almost everyone else in our town and many others around the nation, we had buckets in our shower. It took a bit of experimentation to find the best spots for the buckets where they would catch the maximum water and we would not fall over them. A rectangular dish sat neatly in one of the sink bowls to catch hand-rinsing water, water used to wash vegetables and dregs from the teapot and coffee maker. Water from the washing machine was directed into buckets. So we had three sources of grey water we could reuse.

Some people diverted water from the washing machine into large bins mounted on wheeled trolleys, so they could wheel the water close to the plants and bucket it from the bin. Some found a new use for their wheelie bins as make-do portable tanks. Remember, if you are intending to divert house water into a holding tank, EPA regulations allow grey water to be stored for 24 hours only without treatment.

The next step is to make up a roster. Don't think you will remember from day to day and week to week what plants have had a drink and what haven't. My roster had the days of the week across the top and the water sources down the side. For each day I wrote in the plants that would be watered from each source.

Watering tips

Now that part is organised, you must ensure that every drop of your carefully saved water does its job. You don't want it running off into areas it is not needed. Here are a few suggestions about how to do it.

- Place old buckets with holes in the bottom next to the plant to be watered. Tip your recycled water into the bucket and it will slowly seep through the holes.
- Cut the bottoms off two- or three-litre juice bottles, turn them upside down and push the top opening into the soil, under the mulch. Pour water into the open end. I put many of these around selected plants. Some plants had a bottle on each side. A couple of fruit trees had a bottle on each side and one in the front. You can buy devices that do the same job, but there is

The palms were kept alive by grey water poured into juice bottles.

really no need and it is very satisfying to reuse waste bottles.

- Dig sections of leftover drainage pipe into the soil around a plant as described in Chapter 7.
- In a few cases where I knew the soil structure was good enough that the water would be absorbed quickly and readily without any runoff, water was poured straight onto the ground, over the mulch.
- The mound, moat and mulch method described earlier is ideal for hand watering, ensuring the water stays where you want it.
- Seaweed solution or other liquid fertilisers such as fish emulsion (smelly but wonderful for plants) can be added to the buckets as needed. Seaweed is claimed to increase plants' drought and frost tolerance.

Care with buckets and grey water

Please take care when carrying buckets of water around and do not have them too full. Try to have an equally balanced load in each hand.

Care must be taken about what goes into the grey water. Soaps, detergents, shampoos, conditioners and other toiletries all contain a variety of ingredients which will not necessarily contribute to the health of your soil and plants; in fact, they could be downright damaging. I use shower gel, rather than soap, which is alkaline, and in the washing machine I use laundry balls rather than soap powder. See the grey water section in Chapter 7 for more details. In particular, if you have native plants, which do not need much fertiliser, prefer acid soil and react badly to excess phosphorus and nitrogen, think carefully about which source of grey water you use on them and what has gone into it. I made sure that even the coffee grounds that went into my kitchen waste water were rotated so no one plant got them too often.

Pot plant care

At about the time Stage 3 restrictions came in I decided to move most of the pot plants that were placed around the garden into a central

Not a very pretty sight, but it helped keep the pot plants alive.

area for ease of watering. We erected a length of shadecloth over the area and this doubled the time period between waterings. It did not look pretty, but it did the job. At one stage I watered in some wetting agent, and some seaweed solution was added to the recycled water from time to time. Larger pots had mulch spread on top and upside down drink bottles inserted in them to hold water.

I had several pots containing my favourite tropical hibiscus plants, which I have nurtured through all sorts of conditions for many years. But they do need lots of water. My strategy was that after they flowered I forced them into early dormancy by cutting the stems right back to ground level. They were also put under the shadecloth with the other pot plants so the sun would not bake the soil and damage the dormant root system. This system was a success and the plants regrew.

Forced dormancy

The strategy of forced dormancy described above can be adapted with other plants as well. This is a survival strategy of many deciduous trees. The dropping of leaves in winter helps the plants survive harsh cold conditions in their native habitats. I noticed many of the deciduous trees around town dropping quantities of leaves when they became too heat stressed. Some looked quite dead. But, as soon as we had a couple of falls of rain, new green leaf buds began to appear on some.

In a new garden there is not much scope for pruning plants, but once they have grown a bit you can try this technique. Once a plant begins to look stressed, cut it back by about a third so that there are fewer leaves the plant needs to supply water to and less leaf area through which moisture can be lost to sun and wind. Unless you are philosophical about possible losses, I don't recommend cutting a plant right back to the ground unless you know from experience or observation that it is likely to regrow.

Lawns

I have discussed my feelings about lawns earlier. In short, we need to get over this green lawn obsession. Under drought conditions the lawn must be the first thing to let go. Consider some of the alternatives previously discussed. Almost anything you can think of to do with that lawn area will use less water and be more biodiversity-friendly. You could have a small orchard of fruit trees on dwarfing rootstock that will use far less water than a lawn if watered by drip irrigation or porous hose, can easily be watered by hand during Stage 4 restrictions and is productive to boot. A succulent garden or a bushfood garden in a raised bed on top of the lawn would also work well. Use your imagination. There are more creative and climate-suitable uses for that lawn area.

If you must have the permanently green sward, consider fake grass. There are good synthetics available now. Yes, they do cost, but do your sums. Compare the cost of purchasing and running a mower, the money spent on lawn fertilisers, the time it takes to keep the lawn in good condition, the water it uses, the health and environmental costs of noise and petrol fume pollution and fertiliser runoff contaminating waterways, and, not least, the time, money and work involved in

> ### MORE SUITED TO OUR CLIMATE
>
> The following quote was written in 1839 by John Thompson, Chief Draughtsman, Surveyor General's Department, Sydney:
>
> *'The droughts to which we are so continuously subject render abortive all attempts at maintaining a garden in the English style; and point out to me, that stonework, and terraces, and large shady trees, the characteristics of Hindostanee [sic] gardens, are more suited to our climate than English lawns and flowerbeds.'*

replanting and re-establishing the lawn after every drought.

I described the establishment of my herb lawn earlier. At the height of summer, with no water available to keep it thriving, it was looking very straggly and parts were bare. I had it cut very low and covered it with sugar cane mulch. The area looks neat and I'm sure that most of the herbs will regrow and spread again from roots still in the ground. The mulch will eventually enrich the soil with organic matter and nutrients as it decays, all to the future benefit of the herbs.

Suitable for Australian conditions

As your garden develops in line with suggestions in this chapter, in subsequent droughts there will be fewer plants that need hand-watering. I have a large garden by modern suburban standards, and it was quite young at the time of the record-breaking 2006–07 drought. There are only two of us in the house, we have ingrained water miserliness from 23 years of living with tanks and dams,

Sections of the herb lawn at the end of the drought, mulched to keep it looking neat.

ACTION PLAN

- Prepare the soil so water can permeate easily and be absorbed by organic matter.
- Use mulch.
- Select appropriate species.
- Trial and error, observation and information sharing are vital.
- Protect plants from wind.
- Group plants with similar needs.
- Install rainwater tanks and efficient watering systems.
- Make a plan and a watering roster.
- Be careful with grey water.
- Consider the pot plants.
- Do you really need that lawn?

and the washing machine is a waterwise front-loader that uses about 50 litres per load. I estimate that we had between 350 and 450 litres of recycled water per week to use on the garden, the higher figure only occasionally when we had people staying and there were extra loads in the washing machine. My young garden came through with very few losses and some areas had no extra water.

A drought-proof garden suitable for our Australian conditions is achievable. Not only can we be smarter about the way we garden, our gardens will have a unique Aussie style instead of trying, still, after 220 years, to be copies of styles that suit other countries with vastly different climates.

I will finish this section with a quote from my friend with the sandy soil. She does not like native plants, loves English gardens and cottagey plants. In February 2007, at the height of the drought, when many of her pretty plants growing in sandy soil were dying, she said, *'The longer this goes on the more realistic one becomes'.*

There's nothing like a record-breaking drought to deliver a hefty dose of realism.

APPENDIX 1 PLANTS FOR PARTICULAR PLACES

P = perennial

A = annual

N = native

B = medium to high biodiversity value

S = succulent

Note that these recommendations are quite subjective and based on my own observations and experience. How effective a plant is in attracting pollinators, for example, will depend on many variables specific to your own area. Use these lists as a guide only; your own observations will tell you what works for your garden. In particular, remember that many plants are classified as environmental weeds, or as having the potential to become so, in different areas of Australia. In some cases you might decide that a plant's useful qualities outweigh its potential weediness. If this is the case, be vigilant about preventing garden escapees.

Waterwise plants for small spaces

Anigozanthus spp.	P N B
Kangaroo paw	
Brachyscome multifida	P N B
Cut-leaf daisy	
Celosia spp.	A
Cockscomb	
Chrysocephalum spp.	P N B
Everlasting daisy	
Cotyledon spp.	P S
Craspedia spp.	P N B
Billy buttons	
Crassula spp.	P S
Dampiera diversifolia	P N
Dianella spp.	P N B
Flax lily	
Echeveria spp.	P S
Epacris spp.	P N B
Heath	
Erica spp.	P B
Heath/heather	
Erigeron karvinskianus	P
Mexican daisy	
Goodenia affinis	P N B
Irises spp.	P
Kalanchoe spp.	P S
Lavandula 'Miss Donnington'	P B
Lavandula 'Munstead' (dwarf)	P B
Leschenaultia spp.	P N
Lithodora spp.	P
Lobularia maritima	A
Alyssum	
Lomandra longifolia	P N B
Spiny-headed mat-rush	
Odontospermum maritimum	P B

Patersonia occidentalis	P N
Native iris	
Portulaca grandiflora	A S
Moss rose	
Salvia splendens	P B
(usually treated as annual)	
Scarlet sage	
Sedum spp.	P S
Stonecrop	
Sempervivum spp.	P S
Hens and chickens/houseleek	
Wahlenbergia spp.	P N
Bluebell	
Xerochrysum bracteatum	A N
Strawflower	

Waterwise ground covers

Acacia pravissima	P N B
Ovens wattle, prostrate form	
Arctosis spp.	P
African daisy	
Atriplex semibaccata	P N
Creeping saltbush	
Banksia gardneri	P N B
Banksia spinulosa	P N B
Hairpin banksia	
Brachysema celsianum	P N B
Swan River pea	
Cerastium tomentosum	P
Snow-in-summer	
Chamaemelum nobile	P
Lawn chamomile	

Correa decumbens	P N B
Dampiera diversifolia	P N
Erigeron karvinskianus	P
Mexican daisy	
Gazania spp.	P
Treasure flower	
Grevillea curviloba	P N B
Grevillea gaudi-chaudi	P N B
Grevillea juniperina	P N B
Juniper-leaf grevillea	
Grevillea laurifolia	P N B
Laurel-leaf grevillea	
Kennedia prostrata	P N
Running postman	
Lasiopetalum macrophyllum	P N
Shrubby rusty petals, prostrate form	
Melaleuca lateralis	P N B
Myoporum parvifolium	P N
Creeping boobialla	
Osteospermum spp.	P
Creeping daisy	
Plectranthus argentatus	P N
Silver plectranthus	
Thymus spp.	P
Thyme	

Waterwise shrubs for small to medium gardens

Acacia cognata	P N B
River cascade wattle	
Acacia fimbriata	P N B
Fringed wattle, dwarf form	

Acacia havilandiorum Haviland's wattle	P N B
Acacia lanigera Hairy/woolly wattle	P N B
Alyogyne huegelii Native blue hibiscus	P N
Anigozanthos spp. Kangaroo paws	P N B
Banksia bauri Possum/teddy bear banksia	P N B
Banksia baxteri Baxter's banksia	P N B
Banksia dryandroides	P N B
Banksia robur Large-leaf banksia	P N B
Callistemon citrinus Scarlet bottlebrush	P N B
Callistemon spp.	P N B

Look for many hybrid cultivars such as 'Captain Cook' and 'Little John'.

Coleonema spp. Diosma	P
Dillwynia spp. Eggs and bacon pea	P N
Dryandra spp.	P N B
Eremophila spp. Emu bush	P N B
Grevillea spp.	P N B
Hakea cristata	P N B
Hakea laurina Pincushion hakea	P N B

Hakea sericea	P N B
Isopogon spp.	P N B
Kunzea spp.	P N B
Lavandula spp. Lavender	P B
Leptospermum spp. Tea tree	P N B
Lomandra spp. Mat rush	P N B
Nandina domestica nana Dwarf sacred bamboo	P
Pelargonium peltatum Ivy-leaf geranium	P
Philotheca spp. Wax plant	P N B
Plumbago auriculata Plumbago	P
Rosemarinus officinalis Rosemary	P
Thryptomene spp.	P N
Westringia fructicosa Coastal rosemary	P N B
Xerochrysum bracteatum cvs Everlasting daisies	A N B

Drought survivors in author's garden

Acacia acinacea Gold dust wattle	P N B
Acacia fimbriata Fringed wattle, dwarf form	P N B
Acacia pravisima Ovens wattle, prostrate form	P N B

Acacia verniciflua Varnish wattle	P N B		*Erica Colorans* 'White Delight'	P B
Alyogyne hueglii Native blue hibiscus	P N		*Erigeron karvinskianus* Mexican daisy	P
Arctosis, hybrid African daisy	P		*Eucalyptus caesia* Gungurru, 'Silver Princess'	P N B
Bursaria spinosa Sweet bursaria	P N B		*Eucalyptus citriodora,* *Corymbia citriodora* Lemon-scented gum	P N B
Callistemon 'Captain Cook'	P N B		*Eucalyptus leucoxylon rosea* Red-flowered ironbark	P N B
Callistemon citrinus Scarlet bottlebrush	P N B		*Eutaxia obovata* Egg and bacon plant	P N
Callistemon 'Little John'	P N B		*Grevillea longistyla*	P N B
Callistemon seiberi River bottlebrush	P N B		*Grevillea* 'Pink Pearl'	P N B
Callistemon subulatus	P N B		*Grevillea* 'Red Hook'	P N B
Callistemon viminalis Weeping bottlebrush	P N B		*Grevillea* 'Robyn Gordon'	P N B
Callistemon 'White Anzac'	P N B		*Grevillea sericea* Pink spider flower	P N B
Chrysocephalum apiculatum Common everlasting	P N B		*Hakea marginata*	P N
Coleonema pulchellum aurea Golden diosma	P B		*Indigofera australis* Austral indigo	P N B
Cordyline australis pupurea New Zealand cabbage tree, purple form	P		*Iris*, bearded hybrids	P
Craspedia glauca Billy buttons	P N B		*Iris sibirica* Siberian iris	P
Dianella longifolia Pale flax-lily	P N B		*Lagerstroemia indica* x *L. faurei* Crepe myrtle	P B
Dianella tasmanica Tasman flax-lily	P N B		*Laurus nobilis* Bay tree	P
Echium candicans Pride of Madiera, Tower of jewels	P B		*Lavandula* spp. Lavenders	P B
			Leptospermum 'Cardwell'	P N B

Leptospermum 'Love Affair'	P N B
Leptospermum spectabile Blood red tea tree	P N B
Leucophyta brownii Cushion bush	P N
Melaleuca incana Grey honey myrtle, dwarf form	P N B
Melaleuca lateralis	P N B
Melaleuca laterifolia Robin red breast	P N B
Melaleuca linariifolia Snow-in-summer	P N B
Micromyrtus ciliatus Fringed heath myrtle	P N B
Myoporum parvifolium Creeping boobialla	P N
Nepeta 'Walkers Blue' Catmint	P B
Philotheca myoporoides Australian daphne, Long-leafed waxflower	P N B
Plectranthus argentatus Silver plectranthus	P N B
Rosa spp. Roses	P
Syzigium australe 'Orange Twist' Brush/magenta cherry	P N B
Tanasetum niveum Silver tansy	P
Thryptomene saxicola 'Payne's Hybrid' Payne's thryptomene	P N B
Verticordia plumosa Western Australian feather flower	P N B

Plants for narrow spaces

Acer palmatum 'Sango-Kaku' Japanese maple cultivar	P
Callicarpa spp. Beauty bush	P
Callistemon phoeniceus Lesser bottlebrush	P N B
Callistemon pinifolius Pine-leafed bottlebrush	P N B
Callistemon pityoides Mountain bottlebrush	P N B
Callistemon viminalis Weeping bottlebrush	P N B
Camelia maliflora	P
Camelia purpurea	P
Carpinus betulus 'Columnaris' Common hornbeam cultivar	P
Carpinus betulis 'Fastigiata'	P
Ceanothus spp. Californian lilac	P B
Chamaecyparis lawsoniana Lawson cypress cultivars	P
Chamaedorea seifrizii Bamboo palm	P
Escallonia 'Newports Dwarf' 'Hedge with an Edge'	P
Grevillea banksii Banks's grevillea	P N B
Grevillea bronwenae	P N B
Grevillea olivacea Olive grevillea	P N B

Juniperus communis Common Juniper cultivars	P
Malus 'Ballerina' Ballerina apple	P
Prunus cerasifera 'Oakville Crimson Spire'	P
Prunus × *cistena* Purple-leafed sand cherry	P
Pyrus calleryana capital Flowering pear	P
Ravenala madagascariensis Traveller's palm	P
Rhapis exselsa Lady palm	P
Spirea spp. Bridal wreath	P B
Syringa × *swegiflexa* Lilac	P B
Syringa sweginzowii Lilac	P B
Syzygium australe 'Orange Twist' Brush/magenta cherry	P N B

Decorative trees for small to medium gardens

Acacia leptostachya	P N B
Acacia pendula Boree/weeping myall	P N B
Acacia podalyriifolia Queensland wattle	P N B
Acacia pycantha (Australia's national floral emblem) Golden wattle	P N B

Look for many other small to medium-sized wattles that are highly ornamental.

Acer palmatum Japanese maple, Dissectum group	P
Acer palmatum 'Sango-kaku' Coral bark maple	P
Acer platanoides 'Globosum' Globe-headed maple	P
Albizia julibressen Persian silk tree	P
Alloxylon pinnatum Dorrigo waratah	P N B
Banksia ilicifolia Holly-leaf banksia	P N B
Banksia integrifolia Coast banksia	P N B
Banksia menziesii Firewood banksia	P N B
Callistemon spp.	P N B
Cercis chinensis 'Avondale' Chinese redbud	P
Cornus 'Rutban' Dogwood, hybrid varieties	P
Eucalyptus caesia Gungurru, 'Silver Princess'	P N B
Eucalyptus curtisii Plunkett mallee	P N B
Eucalyptus erythrocorys Red-cap gum	P N B
Eucalyptus macrocarpa	P N B
Eucalyptus pyriformis Dowerin rose	P N B

Eucalyptus rhodantha P N B
Rose mallee

Eucalyptus 'Summer Beauty' P N B

Eucalyptus 'Summer Red' P N B

Fraxinus americana 'Appledell' P
Ash, 'Autumn Applause'™

Hakea eriantha P N B
Tree hakea

Hymenosporum flavum P N B
Native frangipani

Lagerstroemia indica P B
Crepe myrtle, Indian Summer range

Malus florentina P B
Balkan/Italian crabapple

Malus floribunda P B
Japanese flowering crabapple

Malus × *gloroisa* P B

Look for many other small to medium crabapples.

Melaleuca viridiflora P N B
Broad-leafed paperbark, red form

Nuytsia floribunda P N B
Western Australian Christmas tree

Prunus 'Snow Fountains' P
Weeping cherry

Look for many small, often weeping, prunus varieties.

APPENDIX 2 GLOSSARY

Acid soil: Soil with a pH of less than 6, lime is lacking, can be high in organic matter, often occurs in high-rainfall regions.

Alkaline soil: Soil with a pH greater than 8, usually contains a form of lime, often occurs in low-rainfall regions.

Annual: A plant that flowers and fruits in a lifespan of a year or less.

Biennial: A plant with a life cycle of two years. May flower and fruit in both years or only in the second.

Biodiversity: The variety of life forms of a given area – plants, animals and micro-organisms – including the genes they contain and the ecosystems they are a part of.

Biolink: Vegetation corridor that links two or more larger ecosystems of similar vegetation that were historically linked.

Biota: All the life forms in an area – plants, animals and micro-organisms.

Deltoid leaves: Leaves that are more or less triangular in shape, though might have blunt corners.

Ecogardening: The author's own term for a sustainable gardening system using methods that: continuously improve and replenish the soil, avoid using damaging chemical pesticides and fertilisers, are water efficient, incorporate pre-used materials, include native and indigenous plants, avoid environmental weeds and are biodiversity-friendly.

Ecosystem: The living things in a community or geographical area along with the nonliving features with which they interact.

Ecosystem service: Essential services that natural ecosystems provide to humanity, including pollination, food and timber crops, plant-based medicines, nutrient recycling, formation and maintenance of soil, waste disposal, weather regulation, water purification, biomass fuels and neutralisation of toxins.

Friable soil: Soil that is easily crumbled, of loose texture.

Indigenous: Plants or animals that are naturally occurring in a given area.

Loam: Friable soil with a balanced sand, clay and silt content and adequate organic matter. Sandy loam has a higher proportion of sand and clay loam a higher proportion of clay.

Parasitise: The action of one individual (a parasite) feeding on nutrients, fluid or tissue from another (a host), often over a lengthy period and usually weakening but not directly killing it.

Perennial: A plant that lives for three or more years. Trees and shrubs are woody perennials. Bulbs and corms and many other plants with no woody parts above ground are known as herbaceous perennials and usually die down to the ground each winter and regrow from the root mass in spring.

Photosynthesis: The series of chemical reactions that takes place in the green leaves of plants. Chlorophyll in the leaves, energy from the sun, water from the soil and carbon dioxide from the air combine to produce the sugars necessary for plant nutrition.

Pollinator: A creature – insect, bird or small mammal – that assists in transferring pollen grains from an anther to an appropriate stigma so that fertilisation and fruit/seed formation can take place. The anther and stigma might be on the same flower, in different flowers on the one plant or in flowers on separate plants.

Silt: In terms of soil texture silt particles are smaller than sand and larger than clay.

Stomata: Tiny pores on leaves through which water vapour and other gases pass.

Xeroscape garden: Dry area garden which incorporates plants that can survive drought, other than by seed or bulb.

APPENDIX 3 BIBLIOGRAPHY

Australian Water Association, Irrigation Australia Ltd, Nursery and Garden Industry Australia, and Water Services Association of Australia, Sydney: www.smartwatermark.info and www.smartwatermark.org. Information and products for helping to reduce water use around the home, including grey water fixtures.

Australian Weeds Committee, Launceston: www.weeds.org.au. Good site for information on invasive species.

Beattie AJ (1995). *Biodiversity Australia's Living Wealth*. Reed Books, Sydney.

Better Watering Solutions, St Agnes: www.betterwateringsolutions.com.au. A commercially available product measuring both rainfall and evaporation.

Bird Observation and Conservation Australia, Melbourne: www.birdobservers.org.au.

Birds Australia, Melbourne: www.birdsaustralia.com.au.

Buchmann SL and Nabhan GP (1996). *The Forgotten Pollinators*. Shearwater Books, Washington, USA.

Department of Primary Industries, Victoria: www.dpi.vic.gov.au. Includes information on invasive plant species for Victoria.

Flora for Fauna, Nursery and Garden Industries Association and National Heritage Trust: www.floraforfauna.com.au.

French J (1990). *Organic Control of Garden Pests*. Aird Books, Melbourne.

Green Harvest, Maleny: www.greenharvest.com.au. Organic gardening supplies.

Handreck K (1993). *Gardening Down-Under*. CSIRO Publishing, Melbourne.

Handreck, K (2008). *Good Gardens with Less Water*. CSIRO Publishing, Melbourne.

Lanfax Laboratories, Armidale: www.lanfaxlabs.com.au. Soil testing, water testing and information on ingredients in commercial detergents.

Sattler P, Creighton C, Lawson R and Tait J (2002). *Australian Terrestrial Biodiversity Assessment 2002*. Commonwealth of Australia, National Land & Water Resources Audit.

Windust A (2003). *Waterwise House and Garden*. Landlinks Press, Melbourne.

WWF Australia, Sydney: www.wwf.org.au. Includes a section on weeds and pests.

INDEX